Essentials of Cognitive Radio

Do you need to get quickly up-to-speed on cognitive radio? This concise, practical guide presents the key concepts and challenges you need to know about, including issues associated with security, regulation, and designing and building cognitive radios. Written in a descriptive style and using minimum mathematics, complex ideas are made easily understandable, providing you with a perfect introduction to the technology and preparing you to face its many future challenges.

LINDA E. DOYLE is an Associate Professor at Trinity College Dublin, Ireland. She leads a large research group in the Centre for Telecommunications Value-chain Research (CTVR). CTVR focuses on the design of future telecommunications networks and systems, and Professor Doyle's research focuses on wireless networking, cognitive radio, reconfigurable networks, dynamic spectrum access networks and spectrum management techniques. The research involves both theoretical and experimental aspects. Professor Doyle is also vice-chair of the Technical Committee on Cognitive Networks (TCCN) of the IEEE Communications Society.

The Cambridge Wireless Essentials Series

Series Editors
WILLIAM WEBB, *Ofcom, UK*
SUDHIR DIXIT

A series of concise, practical guides for wireless industry professionals.

Martin Cave, Chris Doyle and William Webb, *Essentials of Modern Spectrum Management*
Christopher Haslett, *Essentials of Radio Wave Propagation*
Stephen Wood and Roberto Aiello, *Essentials of UWB*
Christopher Cox, *Essentials of UMTS*
Linda Doyle, *Essentials of Cognitive Radio*

Forthcoming
Steve Methley, *Essentials of Wireless Mesh Networking*
Albert Guiléni Fàbrigas, *Essentials of Error Correction for Wireless Communications*

For further information on any of these titles, the series itself and ordering information see www.cambridge.org/wirelessessentials

Essentials of Cognitive Radio

Linda Doyle
Trinity College, Dublin

CAMBRIDGE
UNIVERSITY PRESS

CAMBRIDGE UNIVERSITY PRESS
Cambridge, New York, Melbourne, Madrid, Cape Town, Singapore, São Paulo, Delhi

Cambridge University Press
The Edinburgh Building, Cambridge CB2 8RU, UK

Published in the United States of America by Cambridge University Press, New York

www.cambridge.org
Information on this title: www.cambridge.org/9780521897709

© Cambridge University Press 2009

First published 2009

Printed in the United Kingdom at the University Press, Cambridge

A catalogue record for this publication is available from the British Library

ISBN 978-0-521-89770-9 hardback

For Philomena Doyle, 1941–2005

Contents

Acknowledgments   *page* xi
List of abbreviations   xiii

1 A cognitive radio world   **1**
  1.1 Introduction   1
  1.2 Brief history and definition   1
  1.3 New spectrum regimes   4
  1.4 Cognitive radio beyond spectrum management   18
  1.5 Application domains   22
  1.6 Conclusions   31
  References   32

2 The essentials – an overview   **33**
  2.1 Introduction   33
  2.2 Setting the scene for understanding cognitive radio 33
  2.3 Building a deeper understanding   38
  2.4 The core essentials   41
  2.5 The other necessities   44
  2.6 A roadmap for the book   45
  References   46

3 Taking action   **47**
  3.1 Introduction   47
  3.2 Understanding the world in which actions take place 48
  3.3 A brief tour of communications systems   57
  3.4 The actions in detail   62
  3.5 Communicating the transmitter configuration details
      to the receiver   81
  3.6 Conclusions   86
  References   87

4 Observing the outside world   **89**
    4.1 Introduction   89
    4.2 The spectrum sensing challenge   92
    4.3 The basic sensing system   96
    4.4 Standalone or non-cooperative spectrum sensing   105
    4.5 Cooperative spectrum sensing   108
    4.6 Getting information from an external source   116
    4.7 Back to the wider observations   120
    4.8 Conclusions   121
    References   122

5 Making decisions   **123**
    5.1 Introduction   123
    5.2 The decision-making process: part 1   123
    5.3 The decision-making process: part 2   139
    5.4 Taking regulations into account when
         making decisions   144
    5.5 The decision-making process: part 3   146
    5.6 Conclusions   152
    References   154

6 Security in cognitive radio   **155**
    6.1 Introduction   155
    6.2 The strength and weakness of being able to observe   157
    6.3 The double-sided coin of collaboration   159
    6.4 Physically tampering with the cognitive radio   161
    6.5 The single points of failure   161
    6.6 Application demands   162
    6.7 An example of security in action   163
    6.8 The silver lining   164
    6.9 Conclusions   165

7 Cognitive radio platforms   **167**
    7.1 Introduction   167
    7.2 A complete cognitive radio system   167

7.3 Cognitive radio platforms: digital hardware   170
7.4 Cognitive radio platforms: the analogue part   182
7.5 Cognitive radio platforms: the other bits   193
7.6 Conclusions   193
References   194

8   Cognitive radio regulation and standardisation   195
8.1 Introduction   195
8.2 Regulatory issues and new spectrum
      management regimes   195
8.3 Cognitive radio applications and regulations   213
8.4 Standards and international activity   214
8.5 Conclusion   220
References   221

9   Conclusions   223
9.1 Introduction   223
9.2 A brief summing up   223
9.3 The future   227

*Appendix A: Developments in the TV
white spaces in the USA*   229
Index   233

Acknowledgments

I have an enormous number of people to thank for helping me with this book. First and foremost I would like to thank all my team in the Centre for Telecommunications Value-chain Research. Many of the team proofread chapters and were always there to discuss topics with me. I would particularly like to thank Tim Forde, Tom Rondeau, Paul Sutton, Ruth-Ann Shields, Jorg Lotze, Rob McAdoo, Keith Nolan, Hicham Lahlou and Colman O'Sullivan for all their reading and for answering my endless questions. I would also like to thank Donal O' Mahony, Pat Hickey, Penny Storey and Irene O' Neill for putting up with my excuses when I spent time writing rather than doing other CTVR work. But most importantly it is because of all of the CTVR team that I have something to write in the first place. And I cannot forget my other colleagues in Trinity, especially Maryann Valiulis, Emma Stokes and John O'Hagan who were always there.

In early 2008, we ran a graduate course in Cognitive Radio and Dynamic Spectrum Access with Virginia Tech. The course was led by Allen Mackenzie, and Luiz Da Silva was also heavily involved. I cannot thank Allen and Luiz enough for what I learned on the course (it is amazing how much you learn when having to teach!). We also had very inspiring guest lectures. It was great to be involved in such a course when writing this book. And I have to mention all my IEEE DySPAN friends from whom I learn so much.

Thanks too to Giuseppe Ruvio, John Keeney and Donal Doyle for taking time to explain some issues of relevance to me.

The original images in this book were designed by Ralph Borland. Thanks to Ralph for taking the time to do this.

And finally I would like to thank Oliver, Tina, Kieran, Colm, Maeve, Margaret, Fiona, Helena, Nicky, Conor, Marie and George for having to listen endlessly to me moaning about this book and for just being generally inspiring and supportive.

Abbreviations

ACG	automatic gain control
ADC	analogue-to-digital converter
AI	artificial intelligence
AM	amplitude modulation
ASIC	application specific integrated circuit
ASIP	application specific instruction set processors
BPSK	binary phase sift keying
CA	Certificate Authority
CDMA	code division multiple access
CPE	consumer premises equipment
CPU	central processing unit
CSMA/CA	carrier sense multiple access/collision avoidance
DAC	digital-to-analogue converter
dB	decibel
DMTF	Desktop Management Task Force
DoD	Department of Defence
DSA	dynamic spectrum access
DSHR	domain specific reconfigurable hardware
DSP	digital signal processor
DTN	disruption/delay tolerant network
DVB-H	digital video broadcast – handheld
EHF	extra high frequency
EIRP	equivalent isotropic radiated power
ELF	extremely low frequency
EMC	electromagnetic compatibility
ETSI	European Telecommunications Standards Institute
EU	European Union
FCC	Federal Communications Commission
FDMA	frequency division multiple access
FFT	fast Fourier transform

FM	frequency modulation
FPGA	field programmable gate array
GPP	general purpose processor
GPS	Global Positioning System
GPU	graphics processing unit
GSM	Global System for Mobile
HF	high frequency
IEEE	Institute of Electrical and Electronics Engineers
IETF	Internet Engineering Task Force
IMD	intermodulation distortion
ISM	industrial, scientific and medical
ITU	International Telecommunication Union
LAN	local area network
LF	low frequency
MAC	media access
MANET	Mobile Ad hoc Network
MF	medium frequency
MIMO	multiple input multiple output
NoC	network on a chip
NRA	National Regulatory Authority
OFDM	orthogonal frequency division multiplexing
PAPR	peak to average power ratio
PDP	policy decision point
PEP	policy enforcement point
PFD	power flux density
PHY	physical
PKI	public key infrastructure
PPE	power processor element
PSK	phase shift keying
QAM	quadrature amplitude modulation
QPSK	quadrature phase shift keying
RF	radio frequency
RTOS	real time operating system
SCF	spectral correlation function
SHF	super high frequency

SNR	signal-to-noise ratio
SoC	system on a chip
SPE	synergistic processor element
TDMA	time division multiple Access
TEM	transverse electromagnetic
UHF	ultra high frequency
UWB	ultra wideband
VHF	very high frequency
VLF	very low frequency
WRC	World Radiocommunication Conference

1   A cognitive radio world

1.1 Introduction

Cognitive radio is a topic of great interest and holds much promise as a technology that will play a strong role in communication systems of the future. This book focuses on the essential elements of cognitive radio technology and regulation. This is a challenging task in that cognitive radio is still very much an emerging technology. There is much debate over its exact definition, its potential role in communication systems, whether cognitive radios should in fact be permitted in the first place and if yes, what the regulatory policies should be. However, while acknowledging the flux in this field, the book aims to identify the core concepts that will remain central to the field irrespective of how precisely it develops. The aim of this first chapter is to briefly define cognitive radio and to then focus on the all important question of why cognitive radios are needed. This chapter therefore motivates all that is to come in the book.

1.2   Brief history and definition

The term cognitive radio was coined by Mitola in an article he wrote with Maguire in 1999 [1]. In that article, Mitola and Maguire describe a cognitive radio as a radio that understands the context in which it finds itself and as a result can tailor the communication process in line with that understanding. Since the coining of the phrase, the term cognitive radio has grown and expanded and has tended to be used in very many ways. With this in mind it is perhaps useful to spend some time defining cognitive radio and place that definition in the context of current radio technologies.

In very simple terms a cognitive radio is a *very smart radio*. Radios have been getting smarter in the last few decades. Current communication systems use radios that can adapt their behaviour in many ways. For example 3G communication devices have the ability to dynamically alter their power output in order to ensure power imbalances, which negatively affect communication, do not arise between different users. Mobile phones can cleverly process the incoming signals they receive in order to mitigate against the various different distortion effects that the signal experiences. WiMAX[1] systems can adapt the characteristics of signals they transmit in order to maintain good throughput and link stability. All this type of functionality is invisible to the user, but the fact remains that the communication systems we use today are able to adapt and change their behaviour in a variety of ways to maintain connectivity in the face of varying conditions and circumstances.

In all of the above examples, the adaptations that occur are well defined and can be anticipated. And the adaptations are triggered by straightforward and well-understood conditions. Let us look at the WiMAX example to see what this means. Modulation is the process by which data is placed on the radio waves for transmission. The order of the modulation scheme gives an indication of how compactly the data is modulated onto the waves. Higher-order schemes get more data through but need good signal conditions to work. Lower-order modulation schemes get less data through but need less good signal conditions. A received signal is typically very good near the basestation and therefore a high-order modulation scheme can be used. However, in areas close to the edge of the range of the WiMAX basestation, the received signal is much poorer. So, the system steps down to a lower-order modulation scheme to maintain the connection quality and link stability. Hence the modulation scheme changes with distance from the basestation as indicated in Figure 1.1. The modulation scheme is known as an adaptive modulation scheme and there is a simple mapping between the quality of the signal at the receiver and the modulation scheme used.

A cognitive radio takes this type of adaptive behaviour and goes much further. By this we mainly mean two things. Firstly the level of adaptivity

---

1 WiMAX is a wireless digital communications system, also known as IEEE 802.16, that is intended for wireless metropolitan area networks.

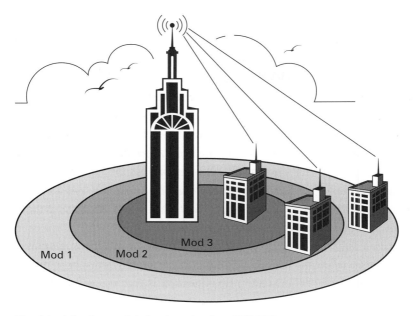

Fig. 1.1. Adaptive modulation in action for a WiMAX system.

is greatly increased and applies to as wide a range of operating parameters as possible, e.g. frequency of operation, power, modulation scheme, antenna beam pattern, battery usage, processor usage, etc. And secondly the adaptation itself can happen in both planned and unplanned ways. The latter can be made possible through the radio recognising patterns of behaviour, learning from reoccurring situations and past experiences and using mechanisms for anticipating future events. With this in mind, a cognitive radio can be defined.

A cognitive radio is a device which has four broad inputs, namely, an understanding of the environment in which it operates, an understanding of the communication requirements of the user(s), an understanding of the regulatory policies which apply to it and an understanding of its own capabilities. In other words a cognitive radio is fully *aware* of the context in which it is operating. A cognitive radio processes the inputs it receives and makes *autonomous decisions* on how to configure itself for the communication tasks at hand. In deciding how to configure itself, the radio attempts to match actions to requirements while at the same time being cognisant of whatever constraints or conflicts (physical, regulatory, etc.) may exist. A cognitive radio has the ability to *learn* from its actions and for this learning to feed into any future reactions it may have. A cognitive radio is made from software and hardware components that can facilitate the wide variety of different configurations it needs to communicate.

This definition, while perhaps not suitable for an elevator pitch, does capture the essential ingredients of a cognitive radio.[2] *How a cognitive radio gets the input it needs, processes the information, decides on how to configure itself, puts its decisions into action and deals with learning is very much the focus of the book.* However, before addressing these issues, the question that needs to be addressed first is, 'Why do we need cognitive radios?' A cognitive radio is undoubtedly more complex than any existing radio and the need for the extra complexity must be justified.

A succinct way of answering this question is that we will be able to do 'new things that we currently cannot do' and we will be able to do 'old things better' with a cognitive radio. The new things that we will be able to do are very much connected with new spectrum management regimes. The old things that we will be able to do better lie mainly in the domain of autonomous organisation and management of increasingly complicated communication systems. And in general it is fair to say that cognitive radios will make communication possible in ever more stressful and challenging circumstances.

1.3 New spectrum regimes

We start by looking at the role of cognitive radio in spectrum management. We do this for two reasons. The first is that historically the motivating applications for cognitive radio have been presented in the context of spectrum management and in particular in relation to *dynamic spectrum access*, a concept which will become clear later in this chapter. The second reason for starting with spectrum management is that many of the wider

---

2 Currently efforts are being made to more tightly define the term. So for example the IEEE 1900.1 Working Group on Terminology and Concepts for Next Generation Radio Systems and Spectrum Management has developed a set of standard definitions.

applications for cognitive radio make use of the dynamic techniques that are fundamental to dynamic spectrum management concepts.

In very general terms, spectrum management involves the process of organising how the spectrum is used and by whom. The key purpose of spectrum management is to maximise the value that society gains from the radio spectrum by allowing as many efficient users as possible while ensuring that the interference between different users remains manageable [2]. New dynamic spectrum management regimes are on the horizon as we move away from the current static approaches and cognitive radio has an enabling role to play.

1.3.1 Current regimes

Historically, the approach adopted by spectrum managers around the world to managing the radio spectrum has been highly prescriptive. Typically, regulators decide on the use of a particular range of frequencies, or frequency band, as well as specifying what services should be delivered in the band, which technologies are permitted in the delivery of the services and who gets to deliver and perhaps use the services. This is referred to as the *administrative approach* to spectrum management. The term *Command and Control* is also often used to describe this approach, or at least to delineate traditional practices which do not necessarily heed market demands, in spectrum management.

The entire radio spectrum is divided into blocks, or bands, of frequencies established for a particular type of service by the process of frequency *allocation*. Frequency allocation is performed on an international and national basis. Broadly speaking, international bodies tend to set out high-level guidance, to which national bodies adhere in setting more detailed policy. At the highest level of management sits the International Telecommunication Union (ITU), a specialised agency of the United Nations. The ITU's International Radio Regulations allocate the spectrum from 9 kHz to over 275 GHz to a range of different uses. In some cases, there are multi-national bodies coordinating across a region. A good example of this is the case of the European Union (EU). At national level typically a national regulatory authority (NRA) manages the day-to-day use of the spectrum, in line with ITU guidelines and for example, in the case of the EU member states, in line with EU policy. Within the broad international frequency allocations, the national regulatory authority can make more specific channel plans. For example, allocations made to the land mobile service can be divided into allotments for business users, public safety users and cellular users, with each group allotted a portion of the band in which to operate. The upshot of this approach is that blocks of the spectrum are set aside for specific activities in each country and the regulator knows what to expect.

The tight control over the *use* of the spectrum is attractive to regulators because with this approach interference can be managed. All communication systems cause a degree of interference. A full description of how interference arises can be found in Chapter 3 but for the moment we can simply think of interference as meaning a distortion of the transmitted waves that can actually prevent the receiver from being able to correctly decipher the incoming signal. In broad terms interference arises when services in neighbouring frequency bands or services in the same frequency bands but in different geographical areas interact with each other in an undesirable way. The administrative approach makes it easier for the regulator to ensure that excessive interference does not occur because the regulator is able to carefully model the interaction between services in neighbouring bands and in different geographical areas and tailor the licence conditions appropriately. This tailoring, for example, manifests itself in the specification of guard bands between services or conditions on maximum power transmission levels. Guard bands are bands of frequencies which are deliberately left free to ensure neighbouring services do not *spill* over into each other. Figure 1.2 captures this concept and shows tightly proscribed neighbouring services, and the associated regulated guard bands, over a range of frequencies.[3] In Figure 1.2 two neighbouring services are defined as neighbours from a frequency perspective. Two neighbouring services can also be defined in a spatial context and it is in this context that the

---

3 This diagram is inspired by a diagram in an Ofcom report which can be downloaded from http://www.ofcom.org.uk/consult/condocs/sur/spectrum/.

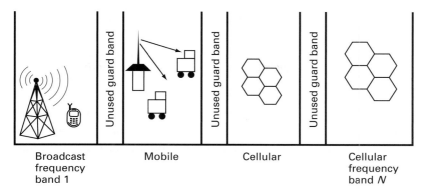

Fig. 1.2. Tightly administered allocations of frequency bands.

power restrictions mentioned above come into play. Chapter 3 will deal with this is more detail.

So, once allocation has taken place, *assignment* is the next stage of the spectrum management process. Assignment, which happens at national level, refers to the final subdivision of the spectrum in which the spectrum is actually assigned to a specific party for use. A wide range of mechanisms exist for making assignments. In cases where there is limited demand for spectrum, something as simple as a first come first served can be used. On the other hand, in cases of high demand, more complex mechanisms are used. For example, in the case of the 3G bands, regulators around the world used beauty contests[4] and auctions to assign the spectrum in the specific bands set aside for these services. Not all spectrum costs money. In the 3G bands just mentioned, large sums of money were handed over for the assignments. On the other hand, TV broadcasters and military users do not pay for spectrum in many countries. The essential point here is not to provide a comprehensive understanding of current spectrum management rules and regulations but rather to underline the fact that centralised decisions, both technical and economic, regarding usage of the spectrum are a key part of the administrative approach to

4 A beauty contest is a merit-based comparative evaluation approach in which interested parties make a submission and are judged and ranked.

spectrum management. Cave *et al.* [2] provide a very good overview of all matters spectrum management.

While the administrative approach has its attractions for regulators and a degree of certainty for incumbents and users, it has many disadvantages. The spectrum planning involved in allocating frequency bands to certain uses and then regulating what equipment can be used to deliver the services can be a slow process and not capable of keeping up with new innovations and technologies. Potential users of spectrum can make proposals for allocations, for example for new communication technologies, but without the allocation being made, matters cannot progress further. So, for example, operators who received licences to provide 3G services are not able to use their allocated spectrum for other services while the 3G market builds. Nor are they easily able to divert 2G spectrum to growing 3G service demands (though this situation is currently being addressed). The frequency allocation process itself forces regulators to 'pick winners'. This can be a very difficult task and it is not sensible or efficient to put the regulator in such a position.[5] There tend to be limited incentives for those who have spectrum to use it efficiently while spectrum is seen as a scarce resource to those in search of it.

However, while on the one hand spectrum is considered scarce, on the other hand there is another story to tell. Figure 1.3 shows a set of spectrum occupancy measurements taken over a particular range of frequencies in a particular city centre location in Dublin City in April 2007. The measurements involved were taken by the Shared Spectrum Company using dedicated equipment and cover a time span which begins at 6 pm on one evening and continues for 40 hours. This image is typical of very many spectrum occupancy measurements[6] in that it exhibits lots of unused spectrum. And in fact figures such as spectrum occupancy levels of as little as 10% are often suggested as giving the real picture of spectrum usage.

---

5 An example of where this has gone wrong is the allocation of spectrum to the ERMES paging system or the TFTS in-flight phone system in Europe. These allocations have resulted in spectrum being unused for over a decade.

6 Shared Spectrum Company have a wide range of such measurements taken in different cities, available on their website http://www.sharedspectrum.com/.

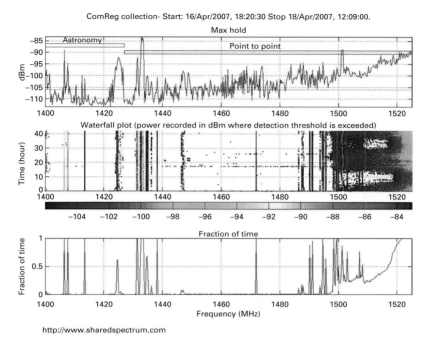

ComReg collection- Start: 16/Apr/2007, 18:20:30 Stop 18/Apr/2007, 12:09:00.

http://www.sharedspectrum.com

Fig. 1.3. Shared Spectrum Company measurements, Dublin, April 2007.

The middle plot of Figure 1.3 is the one that deserves attention. Spectrum occupancy between 1400 and 1520 MHz is shown.[7] As can be seen there are many frequencies in which there are no transmitters present. The temporal nature of the transmissions can also be clearly seen. In particular the transmissions that occur in the 1500 MHz to 1520 MHz range exhibit strong temporal behaviour with significant unoccupied spectrum during two time intervals which in fact correspond to the two nights over which the measurements were taken. The key point to be made is that, despite the fact that all the frequencies in the above bands have been allocated to particular services, all the bands are not fully occupied.

The recognition of the many short-comings in administrative approaches to spectrum management has precipitated a search for new

7 Note that spectrum occupancy in Figure 1.3 is the spectrum occupancy seen by a radio at a given location and another radio elsewhere may get a somewhat different view.

spectrum management regimes and mechanisms. There are many differ-ent innovations on the horizon and there is much discussion and debate and much research as to which approaches are best. What is of interest in this book are the new spectrum management regimes which are rele-vant to cognitive radio. We begin by looking at *dynamic spectrum access* regimes.

1.3.2 Dynamic spectrum access regimes

Measurements such as those in Figure 1.3 and many others have led to the suggestion that it should be possible to access spectrum dynamically. What this means is that radios, rather than being given a static range of frequencies on which to operate, could instead be allowed to use what-ever unused spectrum or *white space* they find free. This is essentially what is meant by *dynamic spectrum access*. The dynamic approach has the potential to make use of what are otherwise wasted resources. The opening up of spectrum in this manner also has the potential to reduce barriers to entry for new ventures as this technique means more spectrum is available for use.

In the first instance the vision for dynamic spectrum access is based on one in which new users 'fill' unoccupied spectrum around existing users. As can be seen from Figure 1.3 some frequencies will be unoccupied for long durations while others are only unoccupied overnight. Others still, not obvious from Figure 1.3, will be much shorter lived and perhaps last minutes and seconds. Needless to say different types of applications may suit longer-term or shorter-term gaps.

In the main, the dynamic spectrum access vision has centred around the concept of spectrum sharing between licensed and unlicensed users. In this case, the holders of licences to the spectrum are known as *pri-mary users*. These licensed users have priority access to the spectrum for which they have a licence and *secondary users* can use the spectrum when the primary user does not need it. The secondary users cease trans-mitting on the return of the primary user. For this to work, the secondary user needs to be able to detect the white space, configure itself to trans-mit in that white space, detect the return of the primary user and then cease transmitting and look for another white space. From the definition of the cognitive radio provided in Section 1.2, it is clear that this falls well within its remit. The secondary user needs cognitive capabilities to both detect the white spaces and to *overlay* its own transmissions on the white space making sure not to interfere with other primary users. Spectrum is therefore shared between the primary and secondary users, with licensed primary users having priority. An *underlay* option also exists. This again is a spectrum sharing regime, except in this case the secondary users share the spectrum by transmitting at the same time as the licensed primary users but at such low powers as to not detrimentally affect the primary users. And, of course, a cognitive radio can refine its methods for accessing the available spectrum by learning the patterns of availability of the white space in cases where patterns exist and perhaps even predict future availability. Figure 1.4 shows a stylised representation of overlay and underlay dynamic spectrum access.

It is perhaps important to note that this *implementation* of dynamic spectrum access tends to be synonymous with the term cognitive radio. For example, the vast majority of research papers in the area of cognitive radio begin by explaining how a cognitive radio will act as a secondary unlicensed user providing either overlay or underlay services to coexist with some primary users. It is worth stressing that, while this is a valid outlook, in the context of this book, we take a much broader view of cognitive radio. *The term dynamic spectrum access simply means that no static assignments of frequencies are made.* Users with static and

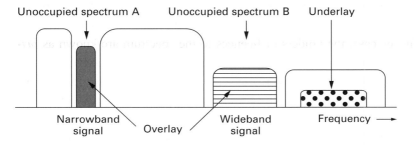

Fig. 1.4. Dynamic spectrum access in action.

non-static frequency assignments can coexist. Whether the radios accessing spectrum dynamically are licensed or unlicensed depends on the regulatory regime in use. And it is possible to envisage a world in which only dynamic spectrum access is in use and no static assignments are made. Furthermore, dynamic spectrum access as defined here is defined as a new means of managing spectrum that is enabled through cognitive radio but can in fact be considered as a generic technique in itself that is used as a means of interference mitigation. We will return to these points repeatedly both in the rest of this chapter and throughout the book.

1.3.3 Technology and service-neutral regimes and trading

Spectrum management regimes based on market mechanisms are seen as key contenders for solving many of the current and future challenges we face. The introduction of auctions as mechanisms for assignment of spectrum was a first step in moving towards more market-driven approaches. However, there is much more to do to free up the highly regulated spectrum world. In general the market-based approaches revolve around the idea of defining spectrum usage rights that would be *exclusively assigned* to a *spectrum consumer*.[8] The term *spectrum property rights* is often used in this context as the spectrum usage rights are likened to those associated with property. In general, with market-based approaches, the aim is to allow uses of the frequency bands to follow market demand as well as to facilitate the buying and selling of those rights in a secondary market. Hence the terms *flexible usage rights* or *flexible spectrum management* or *spectrum liberalisation* are also used.

Recall from the discussion on current spectrum regimes that, in the traditional administrative spectrum management world, spectrum licences are mainly defined in terms of the actual equipment or technologies used to deliver the particular services that are allocated to the particular bands as depicted in Figure 1.2. And the terms of the licences are defined with full knowledge of what equipment and services will be used

---

8 Note we introduce the term spectrum consumer as a general term to denote an actual holder of the specific spectrum usage rights. This is to emphasise a move away from thinking of mobile operators and broadcasters and to think in more general terms.

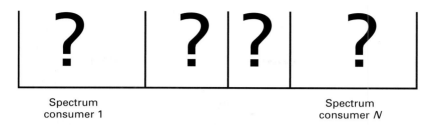

Spectrum
consumer 1

Spectrum
consumer *N*

Fig. 1.5. A service and technology-neutral approach.

in neighbouring bands (with the comfort of guard bands in between). Figure 1.5 is what a market-driven system would look like. The question marks in Figure 1.5 emphasise the fact that any service can be delivered using any technology of choice, in other words they are *technology and service neutral*.[9] Figure 1.5 also attempts to capture 'change of ownership' through emphasising there are no predefined guard bands as it is not possible to define any, given that neighbours can change. The question is still open as to how exactly to define the usage rights of spectrum without specifying services or technologies. The description of these usage rights in terms of frequency, location and time is not enough. Electromagnetic spectrum cannot be confined to neat blocks but, as we learned already, spills over the boundaries. Chapter 8 will discuss some possible solutions. For now it is important to note that some kind of mask or emissions profile will be defined that says to the spectrum consumer 'you will be within your rights provided the profile of your emissions does not deviate from this mask'. The idea of negotiated usage rights also rears its head here as some element of negotiation (call it Coasean[10] Bargaining) may be accommodated within the definition of the usage rights.

9 The meaning of the concept of technology neutrality differs slightly internationally. For example in the EU it tends to be a more narrowly defined term and means that one of a number of technologies and services could be used in a given band. In this book we always use the term in the widest possible sense.
10 Ronald Coase is a Nobel prize winning economist. He is strongly associated with the reform in the policy for allocation of the electromagnetic spectrum, based on his article 'The Federal Communications Commission' (1959) where he criticises spectrum licensing, suggesting property rights as a more efficient method of allocating spectrum to users.

Essentially negotiation allows for deviations from the mask, neighbours permitting.

There is a very strong potential role for cognitive radio in sculpting the emissions profile mask, a task that may no longer be possible through the manual setting of operating parameters of the system because of the interdependency and the complexity of the parameters involved. This is especially true should mask definitions include statistical behaviour over time and space, and even more especially true should some degree of negotiation between neighbours along frequency and spatial boundaries be entertained. The spectrum usage rights approach places the choice in the hands of the spectrum consumer. Spectrum consumers can use simple tactics such as operating self-induced guard bands in order to comply with boundary regulations, which might be the best approach in cases where low-cost technologies are used. Or they can decide to dynamically sculpt and shape the signals they are transmitting. This can be achieved using cognitive radios to maintain the correct power distributions as neighbours change. The learning capabilities of the cognitive radios can also be leveraged to learn which configurations do not work and to remember good responses.

There is also a potential role for cognitive radio in the context of spectrum trading. Spectrum trading is a natural consequence of exclusive usage rights. Spectrum trading is possible on different levels of spectrum granularity and on different time scales. So, for example, there may be initial auctions held by the regulator to assign the available spectrum. These rights may be assigned to spectrum consumers for long time periods and over large bands of frequencies and hence these auctions may take place infrequently. A secondary market will allow spectrum consumers to subsequently trade their rights (change of ownership). Depending on regulations large swathes of spectrum or smaller blocks may be traded or even sublet. Spectrum trading in a general sense is not a main focus of this book; however, where spectrum trading and cognitive radio come together it is. It is possible to envisage scenarios in which the granularity of the spectrum traded (small) and the time scale involved (short) lends itself to the use of smart agents (i.e. cognitive radios). In this scenario bidding for spectrum occurs on an as-needed basis in line with the ebb and flow of user needs.[11] This in fact could be seen as another kind of *dynamic spectrum access* and, taken to its limit, can be thought of as a 'just-in-time spectrum' approach.

The technology and service-neutral approach is very much a *market-based* approach. The regulators don't 'pick the winners'. The market dictates the technologies and services which come to the forefront, and new technologies and novel innovations can be deployed more readily. The facilitation of trading allows spectrum to flow to those who value it most. It is very much the epitome of the opposite of an administrative or Command and Control approach. There are enormous technical, economic, regulatory and political challenges to the ideas outlined here. For the moment our purpose is merely to flag the role of the cognitive radio.

1.3.4 New spectrum commons regimes

At the opposite end of the spectrum[12] to market-based approaches are commons regimes. In a commons regime all users are unlicensed; no users have licences which give them priority access or exclusive usage rights to spectrum. The commons with which we are most familiar is that which operates in the Industrial, Scientific and Medical (ISM) bands. There are a number of bands which have been dedicated ISM bands by the ITU. The most well-known band is 2.400–2.500 GHz (centre frequency 2.450 GHz), the band in which very many wireless local area networks operate. In general the communications equipment must accept any interference generated by other ISM equipment operating in the band.

The ISM bands are very often cited as an example of bands in which there has been much innovation. It is fair to say that a significant amount of progress in wireless communications is due to the technologies that have been developed for these bands. The IEEE 802.11 standard is the wireless local area network standard that is the basis of the wireless networks we use in these bands. The development of this universally

---

11 It should be pointed out that countries such as Australia and New Zealand allow spectrum trading currently in some form. What we are referring to here is trading on a more dynamic and automated basis.
12 Pun intended.

acceptable standard and the ease with which the associated equipment can be deployed owing to the unlicensed status of the bands has led to huge growth in the area over the past decade. Numerous hot spots exist. Novel applications have been developed. And even more importantly the expectation of universally available high-speed wireless access has been created. Many argue that the unlicensed nature of the bands in which the IEEE 802.11 equipment operates has been a huge factor in its success. Hence the interest in more commons-like approaches is not surprising.

The big problem with a commons approach is what is known as the *Tragedy of the Commons*. The tragedy results from the over-use of a free finite resource. Any commons whether it be for unlicensed wireless devices or for the grazing of sheep can suffer in this way. In the case of wireless devices the accumulative interference could potentially become so great that interference-free operation of the radios becomes impossible. In the case of the ISM bands unlicensed devices are constrained to transmit at low powers[13] in order to help mitigate against such a tragedy occurring too easily. Because of the power limitations, the ISM bands are suitable for shorter communication ranges.

So what do we mean by new spectrum commons regimes when a commons already exists? The suggestion of a new spectrum commons approach essentially means taking the ideas of the ISM bands further both by designating more bands as commons and through using far more flexible approaches than setting maximum transmission power levels as a means of control. For example, rather than use power levels as a means of limiting interference, the right to access or use the spectrum could be shared among the users subject to some kind of protocol or *etiquette*. The notion of etiquette needs to be explained. An etiquette typically refers to *conventional rules of personal behaviour observed in the intercourse of polite society*. The same principle is applied in the case of radio behaviour in a commons regime. An etiquette would dictate accepted technical behaviours. For example, an etiquette could allow for bargaining between users, should some users want to transmit over longer

---

13 A set of rules from the Federal Communication Commission (FCC) known as Part 15 (subpart B) govern the behaviour of unlicensed devices.

distances or want greater use of the spectrum at a given time. We need not focus here on what a suitable etiquette for a commons regime would comprise, but simply note that a cognitive radio would lend itself well to etiquette implementation. Etiquettes encompass the notion of understanding of context and situation, and subsequent tailoring of behaviour to suit. Etiquettes tend to be learned over time. A cognitive radio has the level of sophistication needed for management of behaviour on an etiquette basis. Hence the commons as an alternative to Command and Control approaches may necessarily involve cognitive radios, especially if the commons is to be defined in a more sophisticated way than on the basis of power transmission constraints.

The main advantage of the commons is to address the barriers to entry that arise within the current administrative approaches. As an aside it is worth noting that a commons does not equate to access for all but rather to those who conform with the unlicensed protocol (i.e. unlicensed does not mean unregulated). The commons also brings with it a philosophy of openness that can lead to innovation, much as in the case of the already much quoted ISM bands.

1.3.5 The realities of new spectrum regimes

It must be acknowledged that the alternatives to administrative approaches, as outlined here, are described to a certain extent in an ideal manner. There are many technical and regulatory challenges for making them a reality. And there are even greater political challenges. It is perhaps true to say that the successful development of cognitive radio technology will have a very great bearing on whether these alternatives will become a reality. For example, for regulators to permit dynamic spectrum access approaches, conclusive proof must be supplied that cognitive radios can use the empty spectrum while at the same time not unduly interfering with the primary users. To that extent we have somewhat of a chicken and egg scenario as some of the new spectrum management regimes depend on the successful implementation of cognitive radio and the success of cognitive radio may depend on the establishment of new spectrum management regimes.

In any case, the message in the context of new spectrum regimes is that there is a range of roles for cognitive radio. In all of the roles outlined, the cognitive radio uses an understanding of the environment in which it is operating to decide its best course of action, whether that be to use some available white space in the case of dynamic access, to autonomously alter its transmission characteristics in order to not violate its spectrum usage rights, or to control its operating parameters in line with a learned etiquette in the case of the commons.

1.4 Cognitive radio beyond spectrum management

So far we have been talking about cognitive radio in the context of new spectrum management regimes. We now move on to a more general discussion of the role of cognitive radio in communication systems. We look at that role from two perspectives, namely from (1) a system configuration and management perspective and from (2) a user perspective.

1.4.1 System configuration and management perspective

A cognitive radio, or possibly more correctly a network of cognitive radios, can be considered to be a *self-organising* system. The network can understand the context it finds itself in and can configure itself in response to a given set of requirements, in an autonomous fashion. The configuration need not just focus on frequency issues such as dynamic spectrum access and can involve many other features of the network such as power, beam pattern, routing algorithm in use, coding techniques, filtering techniques, etc. However, an element of spectrum management is likely to feature in many configuration processes. Looked at from this self-organising point of view, we can say that any communication application that requires a radio or network of radios to self-organise can justifiably make use of a cognitive radio.

One of the benefits of self-organising systems is *reduced manual configuration effort*. The ability to manage networks without the need for human intervention is of particular importance in areas where operational management costs are prohibitive. This may be the case, for example, in areas which are physically difficult to reach or in developing economies. In the latter, excessive reliance on skilled human intervention may make systems too costly. Or it may be that, as communication systems get ever more complex, manual configuration actually becomes impossible.

Consider current cellular network planning approaches as an example. Cellular networks are planned using a range of tools that generate network plans (e.g. number, positions and operating frequencies of basestations). This stage is typically followed by a variety of processes in which attempts are made to get basestation sites to match the suggested layout and to build and install and configure the basestations as directed. In reality it is never possible to build out a network exactly according to the output of the simulation tools. It may not be possible to get preferred sites, etc. And in any case, no matter how sophisticated the simulation process is, it is not perfect. Hence a stage follows in which engineers change and reconfigure and fine-tune aspects of the network until the network performs at a desired level (i.e. has a certain coverage area, does not drop more than X% calls, etc.). A network of cognitive basestations has the potential to play a role in the changing, fine-tuning and reconfiguring of the network. Any network once installed is not a static entity and needs to be reconfigured, for example, in line with an increasing customer base. Here again self-organising capabilities can come in to play. And on a smaller time scale it is possible to envisage the reconfiguring of networks in order to transfer resources from areas of low demand to high demand as user demands ebb and flow over the day.

While the technological demands of such a sophisticated system may seem very challenging, cognitive radios can be targeted in the shorter term, towards some under-used elements of communication systems. One example that springs to mind here is digital beamforming. Beamforming refers to the manipulation of antenna radiation patterns in order to direct the radiation in a desired direction or directions and away from other directions. However, while such antennas exist we do not see widespread use of these and their potential on a network-wide level has not been fully exploited. An automatic way of doing beamforming is attractive. A way of visualising this is to recall days of an antenna on the roof of a house with one person altering its direction, another watching TV and shouting 'back a bit'... 'over to the left more' in order to get the TV picture as clear as possible. A network of cognitive radios would have the capability to do this on a network-wide level with multiple exchanges of 'back a bit' or 'over to the left' until the best global pattern was obtained.

In terms of allowing existing communication ideas to be exploited the same point can be made in the context of mobile ad hoc networks or MANETs. A mobile ad hoc network can be considered to be a collection of wireless mobile nodes that dynamically form a temporary network on an as-needed basis without the use of any existing network infrastructure. All nodes of the network act as routers and forward received packets to nodes within radio range. The network can grow, reduce in size or fragment in real-time without referencing any central authority. Ad hoc networks have been an enormous research focus in the past decade and thousands of research papers have been written in the field. Yet we do not see any significant level of deployment of these networks. While there are perhaps some business model issues, there are obviously technical challenges that remain unresolved. It may be the case that dynamic spectrum access techniques are needed, not just in the context of making more spectrum available to these networks but in the context of providing the interference mitigation abilities needed to ensure that pockets of the MANET are not interfering with each other. The more general self-organising properties of the cognitive network can also contribute to autonomously configuring the many parameters of the very dynamically behaving ad hoc network in order to deliver a satisfactory performance. Suffice to say, cognitive networks may be the technology that 'make MANETs work'.

Before leaving this section, one final point is worth making. Recall that, in the definition of cognitive radio used at the outset of this chapter, it was stated that, 'In deciding how to configure itself, the radio attempts to match actions to *requirements* while at the same time being cognisant of whatever *constraints* or conflicts (physical, regulatory, etc.) may exist'. In a world where sustainability and energy efficiency are of ever-increasing importance, requirements for meeting energy efficiency goals and constraints around power issues can be set as objectives in the self-configuration process. Which nodes are on and off and, at any time, how information is routed around a network (perhaps flowing through nodes or basestations that are more energy efficient, when possible), schedules for data downloads that can take the pressure off the network and reduce power consumption at peak times, selection of lowest transmission frequencies possible for long-range communications to use less power to get a given range, etc. all may have an effect on the power efficiency of the network. Patterns that lend themselves to most efficient power usage can be recognised and repeated. In other words the ideal of *Green Radio* can become the driving factor behind the configuration process.

1.4.2 User perspective

On a user level, the clever automation of tasks can also play a vital role. The typical user today is faced with ever more choice in terms of communications systems. A wide range of communication systems (2G, 3G, WiFi, WiMAX etc.) and applications (voice, web, mail, location-based services, etc.) exist, and users typically make use of more than one type of system and application. A cognitive radio is one means of providing multiple radios needed to access the various systems in one device. There are other devices that do this, but a cognitive radio can potentially do more than just comply with certain standards such as 3G and WiFi; it can also use mechanisms such as opportunistic spectrum access discussed above. The cognitive radio can also provide the self-organising functionality needed to select between multiple options. For example, choice of communication technology can be automatically made on the basis of cost, bandwidth availability, time constraints, quality of service, energy efficiency or whatever the user deems important. A cognitive radio can orchestrate automatic changeover of technologies, or vertical handover as it is known, when for example coverage issues arise for the technology in use. The self-organising aspects can also manifest themselves through the *personalisation* of the user's experience. Personalisation is about tailoring services and applications to suit the specific needs of the individual user. A cognitive radio can be aware of user position, native language, personal preferences, and can learn user travel routines and habits, and use this information in the personalisation process.

1.5 Application domains

To further the discussion on uses of cognitive radio, it makes sense now to focus on application areas. Because cognitive radio is still in its infancy there is little evidence yet of concrete applications but it is worth emphasising that the applications discussed below under the military, public safety and commercial domain headings are ones which are commonly discussed.

1.5.1 The military domain

The military is a big user of spectrum and of many different types of wireless systems. Communication systems, weapon systems, logistics, radar, sensors, navigation, geolocation and numerous other systems and capabilities are dependent on spectrum access. Typically there is a wide range of heterogeneous communication devices and systems that need to be connected and to inter-operate with each other (e.g. between different sections of a given military force and between different forces in an alliance). There can be land, sea and sky operations in progress. There can be a mixture of centralised and decentralised networks. Many of these systems and capabilities need to be deployed rapidly in unknown and potentially hostile environments. There can be connections that are very temporary (e.g. connectivity delivered via high altitude platforms/aircraft) and longer-term networks that provide more persistent connectivity (e.g. connectivity delivered via semi-permanent ground-based control centres or satellite systems). And there can be varying bandwidth demands, ranging from those needed to deliver real-time high fidelity based video footage to those required for delay-tolerant information. Many of the communication devices and systems are highly mobile, and hence all the issues relating to mobility, interference and connectivity apply. The *netcentric battlefield* is often used to capture the type of picture described here of multiple networks of varying demands and life spans coexisting with each other. It is not a big step to see that a cognitive radio or a network of cognitive radios can potentially be ideal for such an environment. It is a big step to meet the design challenges!

Cognitive radios and cognitive networks can play a role in the initial deployment of networks through facilitating self-configuration of the networks. Any reduced manual configuration effort can have high benefits especially where rapid deployment is necessary.

Cognitive radios and cognitive networks can facilitate the autonomous configuration of different networks to ensure coexistence. In the case where legacy systems exist and there is a mixture of cognitive and non-cognitive networks in place, the cognitive networks can avoid these systems, for example by choosing spectrum bands that are not used by legacy systems. This coexistence is also of huge importance in terms of coalition forces. In this context, cognitive radios can act as bridges between different systems to facilitate inter-operability. A cognitive radio can take an input waveform and translate it into a different output wave-form (by reconfiguring its transmission parameters) and relay it to a different system.

Cognitive radios and cognitive networks can ensure as efficient a use of spectrum as possible. Spectrum can be a scarce resource. The US Department of Defense (DoD) claims that its spectrum requirements are growing by 25% annually while at the same time the temporal and spatial use of the spectrum by its emitters is much less than 1%. And in the battlefield, the available spectrum can vary dramatically with geographic location, and with local communication traffic patterns, both long-term and transitory. There is typically no opportunity to do careful spectrum planning. Techniques which help with making spectrum more readily accessible cannot always be deployed. For example, tall towers which in the commercial world provide line of sight over hilltops, trees and buildings are risky to erect on the battlefield. Dynamic spectrum access techniques lend themselves very well to this kind of scenario, both to make use of whatever spectrum is available as well as to coexist with spectrum users from coalition forces. Spectrum sharing techniques can be used among coalition forces. Certain key systems can, for example, be given primary user status, and opportunistic access of the remaining spectrum can be used by lesser systems. Lower-range communication systems could perhaps use some underlay techniques. The white space that exists can be shared using different approaches, allowing larger amounts to go to systems that have high bandwidth demands.

There is undoubtedly a much wider range of roles for cognitive radios in the military environment that go beyond the simple examples given here. The military domain is a very strong application area. While it is not always easy to get information about military advancements in the public domain, it can be seen from recent DARPA investments that the US military are embracing the cognitive radio approach. And other systems that are currently under development such as the EADS airborne communications node, a highly sophisticated air based system that uses advanced software radio[14] techniques, show indications of heading in that direction.

1.5.2 The public safety domain

A similar discussion on the potential usefulness of cognitive radio applies in the case of public safety. Very many of the same issues arise in both military and public safety scenarios. Issues such as rapid deployment, self-organisation of networks, efficient use of available spectrum, the facilitation of inter-operability through bridging between systems (fire fighters, paramedics, police, ambulance crew, etc.), varying bandwidth demands or the 'moving' of resources to where they are most needed, all arise in the public safety case. As Jesuale and Eydt [3] state in a very comprehensive paper on public safety applications, 'Cognitive radios, once developed and deployed, will allow the disparate frequency bands used by public safety to be "stitched" together and available in a single end-user device. Tunable spectrally adaptive radios and devices that can sense available transmit and receive frequencies, will provide stunning improvements in interoperability and channel capacity for systems in a region and in a State.' The interference environments for public safety applications can be just as difficult to deal with as some of the military environments. For example, high-rise environments of steel and concrete and underground environments can present great difficulties for communication systems. Hence the previous discussion regarding military applications applies here and is not repeated.

One point that was not discussed in the military application section, however, is the notion of *interruptible spectrum*. In many cases around

---

14 A software radio is in simple terms a radio that is made of programmable elements rather than fixed static circuitry. Hence a software radio can be reprogrammed to alter its behaviour.

the world, public safety communication systems have been developed in isolation from commercial infrastructure. The interruptible use of commercial spectrum and systems for public safety would mean that resources could be 'grabbed' when needed (though in some emergencies existing commercial infrastructure can be destroyed). The opposite scenario can also be envisaged. Public safety spectrum tends to be spectrum that is hugely in use or not at all in use as the spectrum is obviously only busy during emergencies. Public safety spectrum could therefore be made available for more commercial oriented services if, and only if, it is possible to quickly reclaim the spectrum on need.

In either of these cases, cognitive radios can play a role. In the case where public services make use of commercial spectrum, the commercial spectrum may come online as a pool of extra resources that can be shared dynamically between public service cognitive radio users. In the second case, commercial cognitive radio users can regularly check permission to use public safety spectrum. Temporary digital certificates can be issued to them, granting access to the public safety spectrum. The temporary certificate can be thought of as a temporary short-lived driving licence that needs to be very regularly revalidated. Alternatively commercial cognitive radio users can treat public services as primary users of public safety spectrum. In this scenario, the secondary commercial users vacate the public spectrum bands on the arrival of the public safety primary users.

Public safety is of high priority, and many would argue that inefficient use of spectrum is a small price to pay for safety. It is only with very robust and reliable cognitive radios that such things as commercial use of public safety spectrum could be contemplated. It should also be noted that, while reaction to the use of cognitive radio tends to be mixed across different regulatory environments, public safety would be seen as one of the more favourable areas of deployment for cognitive radio.

1.5.3 The commercial domain

We now turn our attention towards the commercial world and look at the potential for cognitive radio in the commercial application domain.

There is opportunity for cognitive radios to leverage new spectrum management regimes to gain access to more spectrum for the delivery of whatever wireless services are of interest. There is also potential, just as in the military and public service cases, for using self-organising cognitive radios to enhance current systems as well as to deliver new wireless services. It is possible to imagine a whole range of scenarios that can avail of a cognitive radio. In order to ground the discussion somewhat, the more near to medium-term applications are discussed with the future being left to the final chapter of the book.

The Digital Dividend and the TV white spaces

One of the medium-term possibilities for cognitive radio is in the provision of wireless broadband using TV band frequencies. The background to this needs an explanation. The *Digital Dividend* is the name given to the spectrum that will become free when TV channels switch from analogue to digital operation. This switchover has been mandated for 2012. Analogue TV uses large amounts of spectrum. For example, Ofcom, the UK regulator, states that the five terrestrial television channels that currently broadcast in analogue in Britain (BBC1, BBC2, ITV, Channels 4 and 5) use nearly half of the most valuable bands of spectrum below 1 GHz. Digital broadcasting is roughly six times more efficient than analogue, allowing more channels to be carried across fewer airwaves. The plans for digital switchover will therefore allow for an increase in the efficiency with which the spectrum is used – including the potential for a large amount of spectrum to be released for wholly new services.

To look at the impact of the digital switchover, consider the United States as an example. Here TV stations operate on 6 MHz channels, designated 2 through to 69, namely 54–806 MHz in the VHF and UHF portions of the radio spectrum. To replace the current channels with digital TV channels requires less bandwidth, as mentioned above, with all the digital TV stations contained within channels 2–51. The portion of the spectrum channels that covers 52 through to 69, namely 698--806 MHz, has been reallocated to other services. In the channels occupied by the digital TV stations, i.e. channels 2–51, in any given geographical area, not all available channels will be used. The reason for this is that certain physical spacing is needed between each digital station to ensure interference won't occur. In addition, in some areas, not all of the channels that could be used by TV stations will be used. In most rural areas, there are more empty channels than occupied channels. Even in urban areas, a substantial amount of spectrum could be made available for wireless broadband.

In all countries the same issue will arise once digital switchover has happened, i.e. there will be an amount of white space that will sit there unused in the current regime. As an example, Figure 1.6 shows TV channel occupancy in San Francisco *after* switchover and, as can be seen, a significant amount of white space exists. The measurement is taken from the New American Foundation article on 'Measuring the TV "White Space" Available for Unlicensed Wireless Broadband'.[15] Similar pictures exist for other areas; even more space is available in rural regions.

As stated at the outset of this section, one of the possibilities for cognitive radio is in the provision of wireless broadband using TV band frequencies. Cognitive radio can overlay its transmissions on the TV

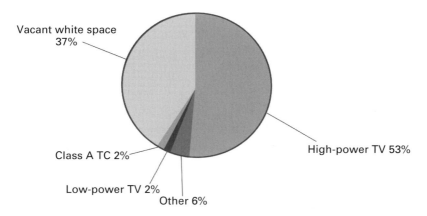

Fig. 1.6. TV white space in the San Francisco area after the switchover to digital TV.

15 http://www.newamerica.net/publications/policy

white spaces. To do this the cognitive radio must be able to accurately detect the presence of the incumbent users, i.e. the TV stations, and as it turns out in this case also wireless microphones, and to subsequently shape its transmissions to fit in the white spaces. This kind of functionality is well within the grasp of cognitive radio.

In December 2007, the UK regulator Ofcom opened up the way for this to become a reality in the UK. Ofcom is proposing to allow cognitive radio to use what they call 'interleaved spectrum' (otherwise known as the TV white spaces) provided that it can be shown that this does not cause interference to other spectrum users. The TV white space or interleaved spectrum is the first strong and realistic commercial opening for the use of cognitive radio. In November 2008 the FCC released a report defining rules for unlicensed use of the TV bands (see Appendix A for details).

Spectrum underlay for short-range communication systems
The TV white space is an example of where commercial overlay systems can be exploited. Ultra wideband (UWB) technology is seen as a means of delivering on the underlay front. A UWB radio, as the name would suggest, is a radio that transmits a signal over a very wide band, at low power. This idea is captured in Figure 1.7. The UWB signal is much wider

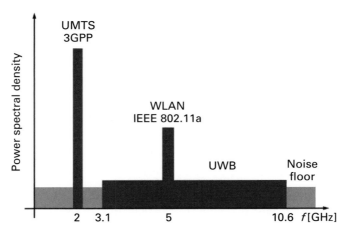

Fig. 1.7. UWB signals are very wide.

than any of the other signals and just above the noise floor.[16] Therefore UWB transmissions could coexist with more powerful signals and not interfere in a significant manner. UWB transmitters can transmit high data rates and would tend to be used for shorter distances, in the range of metres.

UWB has been a controversial technology. The use of UWB has been challenged because of its potential to cause interference, especially to low-powered receivers such as GPS (Global Positioning Systems) receivers. In response to this a static UWB mask was agreed that limits UWB emissions in certain sensitive bands. A cognitive version of a UWB radio would be able to determine when and where it should transmit in a much more dynamic fashion and would perhaps overcome some of the issues relating to the use of UWB.

Spectrum underlay for short-range communications make sense beyond the UWB case. Any spectrum has the potential to be used in a low-powered way for short-distance transfer of high-speed data provided the system can get a sense of potential interference.

Cognitive radio in the cellular world

The cellular world also provides potential for cognitive radio applications. There are a number of different ways in which cognitive radio can play a role.

In Section 1.3 spectrum trading was discussed in what is perhaps an ideal and more long-term manner. Spectrum trading or some form thereof is possible in the shorter term within closed groups. One such closed group is a group of cellular operators. The relevant trading concepts revolve in the main around some kind of *spectrum pooling*. To pool means *to put (resources) into a common stock or fund; to share in common, to combine for the common benefit*. In the case of spectrum pooling, all available spectrum is put into a common fund, redistributed using a suitable mechanism, used for a certain time span, put back into the pool and redistributed all over again. The suggested distribution mechanism, in these scenarios, tends to be some kind of auction mechanism. For

16 The noise floor is the general level of background noise that exists.

example, those eligible to use the resources of the pool place bids at the beginning of each distribution phase and the highest bidders win. This mechanism can be seen as an alternative to statically over-provisioning at the outset and allowing some way of reacting dynamically to user demands. A cognitive radio has the potential to act for the user (as an autonomous agent), making bids for pooled spectrum.

There are many different possible flavours of spectrum pooling in which the cognitive radio can participate. A group of operators could for example decide to collectively pool *all* their spectrum resources and subsequently bid for use of the spectrum during each distribution phase. The income generated in this case is distributed among all operators in proportion to the spectrum each operator put in the pool. Operators who put in more than they bid for may make a surplus, and operators who bid for more than they put in have a cost during that phase. Alternatively, a group of operators could for example decide to collectively pool *part* of their individual spectrum resources. In this scenario, each operator has a base spectrum allowance, which can be augmented from the common pool. Again here the income generated is distributed among all operators in proportion to the spectrum each operator put in the pool. Another alternative still is to have a spectrum pool operated by a *third party*. Operators can bid as in the previous two cases for access to the pool of resources. However, in this case the third party receives the revenue for the spectrum. There is much discussion of the notion of third party spectrum brokers who would also operate in such environments.

This is a form of dynamic spectrum trading, albeit in a limited sense. In the scenarios described, there is a limited amount of spectrum available and there are a limited number of entities who are entitled to bid for spectrum, and 'ownership' of licences does not actually change hands. Some flavours of spectrum pooling have been suggested that open the pooled resource to a wider audience such as unlicensed users as well as the licensed operators. Other flavours of pooling have been suggested that are limited to within the bounds of one operator, and in this case are used as a means of better distributing resources among different entities under the control of that one operator.

Trading can also be conceptualised in the context of roaming. Currently users roam on other networks when abroad. There is no technical impediment to roaming from network to network at home, rather it is the preferred mode of operation for operators to 'own' users. In reality users should be able to access the best option for communication (in terms of price or otherwise) at a given instance and move to the network of choice (very much in the manner that it is possible to manually select a preferred network while abroad, except in a more automated fashion).

Short-term white spaces and disruption-tolerant networking

The TV white spaces will tend to exist in a given area for lengthy periods of time. White spaces of shorter duration of minutes and seconds can also be dynamically accessed. The obvious question is what kind of applications suit these kinds of white space? One answer is that, provided cognitive radios can seamlessly move from one white space to the next, without interruption, any application can work. This capability, however, may be a certain distance in the future. Another answer is to look towards a *delay-tolerant network* or *disruption-tolerant network* and the associate applications.

A disruption-tolerant network is designed to cope with temporary or intermittent communications problems or limitations. The concept embraces the notion of occasionally connected networks and applications that run on such networks need to be delay or disruption tolerant. A voice call could not stand such disruption but a file transfer could. Machine to machine communications might perhaps suit a data transmission network (DTN), especially in the context of large corporate data transfers that can happen out of hours or overnight. There seems to be a natural match between the concepts of disruption-tolerant networking and intermittent white space availability. A delay-tolerant network could make use of the two overnight white spaces that were identified in Figure 1.3 for example for machine to machine communications. There are also many examples of disruption- and delay-tolerant applications in environmental sensing and other areas.

1.6 Conclusions

It could be argued that a great deal of what has been discussed in this chapter can be read as speculation as it is not yet clear how much the cognitive radio will truly be able to deliver. On the other hand it can be argued that a great deal of what has been discussed in this chapter is possible and the chapter is simply establishing a road map. Either way, the message from this chapter is that cognitive radio is a technology that has much promise. We need cognitive radios to be able to deliver new ways of managing the spectrum. We need cognitive radios to be able to efficiently and effectively manage the very complex and multi-faceted communication systems that exist today and to underpin the more dynamic self-organising systems of the future. And there is much potential for exploitation in the military, public safety and commercial domains.

We end the first chapter with a quotation from an article on cognitive dynamic systems by Haykin [4], in which he talks about: 'the design and development of a new generation of wireless dynamic systems exemplified by cognitive radio and cognitive radar, with efficiency, effectiveness, and robustness as the hallmarks of performance'. In doing this, Haykin more or less points the way to the next generation of wireless systems.

Perform ance parameters

References

1. J. Mitola and G. Maguire, Cognitive radio: Making software radios more personal, *IEEE Personal Communications*, **37**:10 (1999), 13–18.
2. M. Cave, C. Doyle and W. Webb, *Essentials of Modern Spectrum Management*, Cambridge Wireless Essentials Series. Cambridge University Press, 2007.
3. N. Jesuale and B. C. Eydt, A policy proposal to enable cognitive radio for public safety and industry in the land mobile radio bands, in 2nd *IEEE International Symposium on New Frontiers in Dynamic Spectrum Access Networks*, 2007. 17–20 April, 66–77.
4. S. Haykin, Cognitive dynamic systems, *Proceedings of the IEEE*, **94**:11 (2006), 1910–11.

2 The essentials – an overview

2.1 Introduction

The first chapter of this book focused on the application areas that will drive cognitive radio technology. This chapter acts as a bridge to the remainder of the book. It seeks to provide the reader with a broad sense of *all* that is involved in cognitive radio technology. In order to do this we go to the heart of the cognitive radio but not at first using technology as an example. Instead we step back and take a look at how decisions are made in a more abstract manner before returning to the radio world. The final part of the chapter provides a roadmap for the rest of the book.

2.2 Setting the scene for understanding cognitive radio

The first question to think about is: how do we make decisions? How do we reason and come to conclusions? We begin this discussion by looking at a simple example.

2.2.1 The lone radio

Scenario 1: I am about to go out and must decide whether I should take an umbrella with me or not. The umbrella is heavy and cumbersome and, while I don't want to get wet, I don't want to take the umbrella with me if it is not necessary.

In this example two actions are possible, namely *take umbrella* or *don't take umbrella*. I need to determine how likely it is to rain in order to decide whether to take the umbrella or not. I can arrive at my decision in a number of different ways:

1. I look out of the window, see that it is bright and sunny and decide not to take an umbrella with me.
2. I look out of the window and, while it is not currently raining, I see very dark clouds in the sky, so I decide that there is a good risk of rain and I take an umbrella with me.
3. I look out of the window, see that it is bright and sunny but recall that at this time of year there are always sudden showers in the afternoon and therefore I decide to take an umbrella with me.
4. I listen to the weather forecast and, even though there is only a very slight chance of rain, I take an umbrella as I recently have had a very bad cold.

While the example scenario is simple, quite complex cognitive functionality can be brought to bear. The four cases represent decision-making processes of varying levels of complexity that use varying levels of information. In case 1, observations regarding the current state of the weather are used to inform the decision-making process. There is a simple mapping between what is observed and what decision is made. *I see rain hence I take an umbrella. I see sunshine, hence I don't.* In case 2 some predictive analysis is performed. In this case clouds can signal the potential of rain. The actual characteristics of the clouds can reveal the likelihood of rainfall. So light grey clouds may just mean it will be a grey day whereas dark clouds may indicate rain. Hence there is a more complex mapping involved here that includes some notion of a greyness threshold that must be exceeded before I take the umbrella. In case 3, past experience and learning comes into play. In this case, more information is used, such as the time of day and the season of the year, to be able to bring learning to bear on the decision. In case 4 I let the cost of getting the decision wrong dominate my choice. In other words getting another cold is worse than the inconvenience of carrying the umbrella. Case 4 is also another example of where external information (i.e. the weather forecast) comes into play.

This umbrella example highlights a very import point. If we look at what is going on here we see that observations are made, feed into a decision-making process, and an action results. This is what we call an **observe**, **decide**, **act** cycle. The decision-making process can be very simple or can involve more complex processes with past experience or future probabilities and risk analysis of various events coming into play. In other words there is a whole range of ways of coming to a decision. The complexity of the process will tend to dictate the kind of observations that are needed; in the umbrella scenario, weather, time of day and season, characteristics of clouds, state of well-being of the individual are some of the various observations that can be used. Some observations are directly observable while others need an external source/database (calendar) or other piece of equipment (clock) for further information. Irrespective of the complexity, the 'observe, decide, act' cycle captures the essentials of the process.

This process is mirrored exactly in a cognitive radio. We can see this by recalling the definition of a cognitive radio as set out in Chapter 1 and repeated in Figure 2.1. The 'observe, decide and act' cycle is distinguishable in this definition. The understanding of the radio environment, the user requirements, the existing constraints, etc. is brought about through observation. The decision process uses the observations as inputs and produces an action or a set of actions as outputs. The particular type of actions all relate to the configuration of the cognitive radio. Hence we can say that the cyclical process of observe, decide and act represents the core functionality of a cognitive radio.

If we step back a bit, it could be argued that the 'observe, decide, act' cycle can be seen as representative of any adaptive communication process, such as the adaptive modulation schemes used in the WiMAX

Observe	A cognitive radio is a device which has four broad inputs, namely, an understanding of the environment in which it operates, an understanding of the communication requirements of the user(s), an understanding of the network and regulatory policies which apply to it and an understanding of its own capabilities. In other words a cognitive radio is aware of the context in which it is operating.
Decide	A cognitive radio processes the inputs it receives and makes autonomous decisions on how to configure itself for the communication tasks at hand. In deciding how to configure itself, the radio attempts to match actions to requirements while at the same time being cognisant of what ever constraints or conflicts (physical, regulatory, etc.) that may exist. A cognitive radio has the ability to learn from its actions and for this learning to feed into any future reactions it may have.
Act	A cognitive radio is made from software and hardware components that can facilitate the wide variety of different configurations it needs to communicate.

Fig. 2.1. Seeing the 'observe, decide and act' cycle in the definition of cognitive radio.

systems described in Chapter 1. Cognitive radios, as stressed in the opening chapter, are radios that will facilitate much greater degrees of flexibility, adaptability and dynamism than is currently the case. If we go back to the umbrella example we could say that the adaptive WiMAX system is much more like case 1 in which a simple mapping between observed circumstances and action takes place (i.e. see rain, take umbrella; don't see rain, don't take umbrella). A cognitive radio embraces the more sophisticated elements of decision-making that go beyond the simple mapping between observation and action to more complex predictive analysis and learning, as in cases 3 and 4. The cognitive label therefore is synonymous with communication systems that are highly flexible, highly responsive and extremely dynamic by virtue of being able to make the kind of complex decisions that call on a variety of approaches to decision-making.

The simple 'observe, decide, act' cycle is depicted in Figure 2.2. The representation is useful in that it emphasises that observations are made in the real world and actions are taken in the real world. It is worth noting that in general more complex versions of the 'observe, decide and act'

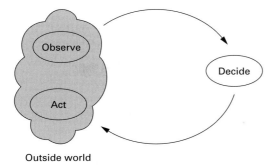

Fig. 2.2. The 'observe, decide and act' cycle. This forms the basis of the book.

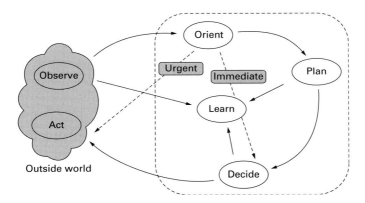

Fig. 2.3. The cognition cycle based on the cognition cycle in Mitola and Maguire [1].

cycle tend to be used to represent cognitive radio functionality. These more complex versions, such as the one used by Mitola and Maguire [1], depicted in Figure 2.3, include orient and plan stages and explicitly mention the learn stage. Orient tends to refer to the process of determining whether urgent action is needed or more long-term planning is involved. Plan involves making decisions for the longer term. In this book we simply summarise all that is involved in decision-making, whether that involves determining how quickly a reaction is needed or learning and planning as the *decide* stage of the process, and therefore, we work with Figure 2.2.

2.3 Building a deeper understanding

The discussion thus far has captured some of the most important ideas that are central to an understanding of a cognitive radio and that are needed in the rest of the book. However, what has not been captured is the fact that a cognitive radio is *not* a standalone entity but rather one element of a network. We consider a second example to explore this point.

2.3.1 The radio in the group

Scenario 2: A group of friends and I have bought a time-share apartment in a holiday resort on a Greek island and we want to make out a schedule for use of the apartment that satisfies as many group members as possible.

In this second scenario a large number of possible schedules can be created to suit the needs of the group members. The decision in this case is for the group as a whole. There are a number of possible approaches that can be used to decide on the schedule to use:

1. We can elect a leader who has sole responsibility to choose the schedule.
2. We can elect a leader who consults all group members and on the basis of the consultation creates some kind of schedule.
3. We can come to some kind of consensus through exchanging ideas and opinions until one clear schedule emerges.
4. We can each individually decide to do our own thing and use the time share whenever it suits and hope a conflict does not arise.

In cases 1 and 2 the decisions are formed in some kind of centralised manner. In case 2, there are many ways of consulting individuals and getting their input, from the superficial to the deep. In case 3 an effort is made to reach consensus without the need for central control. In this case it is easy to imagine multiple parallel conversations taking place between members of the group, and individuals' preferences changing and possibly oscillating to and fro from certain dates, as these conversations evolve, until a collectively agreed schedule is found. In case 4 we see selfish behaviour dominating. With luck there will be no clashes but the risk of two or more people turning up at the same time is obviously great. All four cases illustrated here collectively show how individuals can play a greater or lesser role in the decision-making process, depending on the mechanism used.

The **observe, decide, act** cycle can be seen here too. The observations in this case consist of the individual observations (preferences) of the group members which feed into the larger decision-making process (except in the case of the elected leader with sole responsibility). The decision-making process in this case is some kind of process that facilitates the *combining or collating or fusing* of the individual preferences and the generating of the most suitable schedule. And the action in this case is the putting in place of the schedule for use of the holiday apartment.

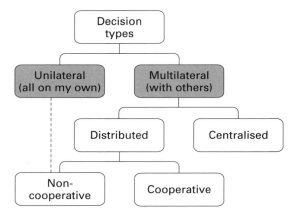

Fig. 2.4. A taxonomy of decision types covering both the umbrella and apartment time-share examples.

If we now think of both the umbrella and time-share analogies, we can come up with a *taxonomy of decision-making processes*. Figure 2.4 attempts to capture this. The more formal title of unilateral decision is given to the decision made in a standalone fashion (umbrella) while the term multilateral decision captures the group-based one (time share). The group decisions can be made in some kind of centralised manner (election of a leader as in cases 1 and 2) or can be made in a distributed fashion (parallel conversations as in case 3). The distributed decision can be cooperative when a consensus emerges or uncooperative (do own thing as in case 4).

This taxonomy maps directly to cognitive radio. The centralised decision-making examples correspond to situations in which infrastructure such as basestations and access points exist and the decentralised mechanisms correspond to entities like ad hoc networks in which there is no central point of authority. In the latter networks, individual radios can cooperate to come to a collective decision or act in a selfish manner and just suit themselves. An individual node in either of the networks can make a unilateral decision.

2.4 The core essentials

So armed with a general understanding of observe, decide and act, we now set out some more of the essential issues and we stay firmly back in the radio domain. The logical place to start in the 'observe, decide and act' cycle is actually at the act end. The reason for this is that we need to know what actions are of interest, and indeed possible, before we can understand what observations need to be made and what decisions need to be taken. The next three chapters of the book will deal with 'Taking action', 'Making observations' and 'Making decisions' in turn and these can be considered to be the central chapters of the book. Here we preview the main issues involved to further build our broad understanding before going into detail in subsequent chapters.

2.4.1 Taking action

It has been suggested that a useful way of thinking about cognitive radio is to think in terms of *meters* and *knobs*. Meters can be read to give an indication of the state of some quantity, and knobs can be set to give a certain kind of behaviour or performance. Consider a mobile phone. The meters on a mobile phone show the state of the signal in bars and show how much battery remains in terms of a battery icon. There are many knobs that can be set. As a user, you can set ring tone and screen savers and there are all sorts of other knobs that are invisible to you, such as power levels or frequency of operation that can be set by the phone. As a user you can observe the meters and respond by setting a knob to a different value. So for example you can observe low battery and set Bluetooth connectivity to off as Bluetooth drains the battery particularly quickly. This kind of process goes on behind the scene as well, with the innards of the phone, for example, using various observations about the received signal to trigger certain settings of the system. When we extend this to a cognitive radio we imagine many more meters and many more knobs to play with. In the language of meters and knobs, the meters correspond to the observe part of the cycle and the knobs to the act part of the cycle. The decide is simply the bit in between!

With this in mind, taking action is all about setting the various knobs of the cognitive radio to create the kind of performance you want. The number of knobs in a radio depends on the design of the radio. Cognitive radios tend to be designed with as many knobs as possible or at least with the aim of creating knobs that will allow the radio behaviour to be altered as effectively as possible to deliver the kind of applications that were discussed in Chapter 1. To think about the range of knobs it is useful to think of the different layers of a communication system.[1]

There will be a range of knobs at the physical or PHY layer that will allow physical characteristics of the transmitted waveform to be changed. The signal shape, frequency, bandwidth, duration, the amount of information it bears, the direction it is travelling in and the power at which it is transmitted are examples of the knobs that can be set. There are possibility for knobs at the Media Access or MAC layer. The coding used to make the signal robust to interference, the way information in the signal is organised and broken up, the way the signal shares the wireless medium with other signals are among the knobs that can be set. The same argument can be made at the network layer. Knobs related to the routing of the data can be set; which routing schemes are used, how routing information is stored and how much is stored can all be variable parameters. There can be more general knobs used to manage the battery in the radio, and knobs related to how the processor is used. This list is by no means exhaustive but gives merely a flavour of the possibilities.

The different applications described in Chapter 1 will each necessitate the setting of one or more knobs (one or more actions). Chapter 3 will focus on actions in detail and give a sense of all that is possible in a cognitive radio. To understand which actions are possible, the consequences of the actions of the cognitive radio need to be understood. This

---

1 A node of a network, such as a radio in a wireless network, is often defined in terms of layers. A layered system can be designed in a modular fashion and hence its attraction. A layer is a collection of related functions that provides services to the layer above it and receives service from the layer below it. The physical or PHY layer defines the physical means of sending data over the wireless link. The Media Access, or MAC layer, frames the data to be sent, detects and corrects errors in data and manages the sharing of wireless medium among different users. The network layer routes data around the wireless network and can provide flow and congestion control. Other layers can sit on top of these but for the purposes of this book these are the main three of interest.

essentially means understanding the interference a cognitive radio can cause to others. This topic is a major concern of Chapter 3.

2.4.2 Making observations

To choose the settings for the knobs, some clues and pointers are needed. From our discussion thus far we understand that meters guide our way. The term meter, while useful, does not fully give the sense of all the inputs that a cognitive radio can use to help choose the settings for the knobs. We return to our definition of cognitive to get the fuller sense of the observations needed. Of interest is that section of the definition which states that 'A cognitive radio is a device which has four broad inputs, namely, an understanding of the environment in which it operates, an understanding of the communication requirements of the user(s), an understanding of the regulatory policies which apply to it and an understanding of its own capabilities.' If we take each of the four inputs we can begin to understand the observations that might arise.

To understand the environment in which it operates, a cognitive radio may need to make observations about the quality of the received signal, whether other signals are being transmitted, the number and location of the other signals being transmitted, the nature of the radio environment and how it will impact on any signals transmitted by the radio. The cognitive radio may need to glean information about the size of the network it is operating in, the number of neighbouring nodes it has, the level of mobility of itself and other nodes, and the traffic levels in the network. Its own location, the time of day, the temperature and other characteristics of the physical environment may also be of interest. It should be flagged at this point that a cognitive radio may not always be capable of making an observation on its own and hence either have to enlist the 'help of friends' or indeed turn to some external source.

To understand the communication requirements of the user, the radio may need to make observations about what kind of applications the user is running or have some means of gaining information about user preferences through direct input on the users' part or through making observations of user behaviour over an extended time period. To understand the regulatory policies that are in play, a cognitive radio needs to know where it is and what policies apply in that jurisdiction. These kind of observations will in general need to be supplied by some external authority. To understand its own capabilities the radio will need to be capable of observing its own current operating parameters as well as understanding the knobs it has available to it and the range of settings that are possible.

The applications described in Chapter 1 will necessitate making one or more of these observations. Chapter 4 will describe the various techniques a cognitive radio can use to make observations. These techniques all involve either using the functionality of the cognitive radio to make observations and/or obtaining observations and information from outside entities. Both approaches are important concerns in Chapter 4.

2.4.3 Making decisions

The decide stage of the process then uses the observations as input to the decision-making process in order to determine how the knobs of the cognitive radio should be set. The applications described in Chapter 1 will necessitate the making of one or more decisions (i.e. deciding on the setting of one or more knobs). The umbrella and the apartment time-share examples will be referred to throughout, as most decisions tend to be either like the umbrella or like the time share; they will either be unilateral or multilateral in nature. Chapter 5 describes different mechanisms for making both types of decisions. And as underscored in the expanded cognitive cycle of Figure 2.3, learning can be brought to bear on the decision-making process. Learning is an important aspect of Chapter 5.

2.5 The other necessities

While the core of the cognitive radio is embodied in the 'observe, decide and act' cycle, three other issues are key. Firstly the cognitive radio must be made secure. There are many unique security threats that arise for cognitive radio that have not been issues for more traditional technology. Chapter 6 deals with this topic.

Secondly there are many challenges associated with building a cognitive radio. These challenges cross the digital and analogue domains.

The type of functionality demanded by a cognitive radio will stress the analogue radio frequency (RF) frontend of the system as well as demand sophisticated digital process. Chapter 7 therefore focuses on the all-important topic of cognitive radio platforms.

And thirdly, and of no lesser importance, are the regulatory challenges and the associated challenges of policing and enforcement. Without appropriate regulations and appropriate policing, cognitive radios have no future. Hence Chapter 8 focuses on these topics.

As cognitive radio is still in its infancy many of the issues relating to securing, building and regulating are open. By the time this book is published there will no doubt be new advances. With this in mind the chapters aim to establish the key issues. And while they include example approaches and algorithms and suggestions for implementation, the essential point is to not get bogged down in any one specific example but instead to embrace the general concept.

2.6 A roadmap for the book

This final section summarises the topics mentioned already and provides a roadmap through the book, couched in terms of the discussion that has

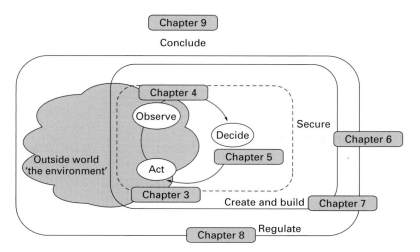

Fig. 2.5. A map of the book based around the all-important 'observe, decide and act' cycle.

taken place in this chapter. Figure 2.5 is a graphical depiction of that roadmap.

Chapter 3 deals with the issue of taking action and the implications for actions in the outside world. Chapter 4 deals with how observations are made. Chapter 5 focuses on the decision-making process. Chapter 6 focuses on the security issues. Chapter 7 takes a more practical approach and is concerned with the various options that exist for designing and building cognitive radios hence, encapsulating the functionality emphasised in Chapter 2 in a real system. Chapter 8 discusses the implications for regulation of cognitive radios and the technical demands that may arise. Finally Chapter 9 concludes by focusing on the future challenges for cognitive radio.

References

1. J. Mitola and G. Maguire, Cognitive radio: Making software radio's more personal, *IEEE Personal Communications*, **37**:10 (1999), 13–18.

3   Taking action

3.1 Introduction

We are by now well acquainted with the 'observe, decide, act' cycle. This chapter is the first of three core chapters that relates to the cycle and focuses on the act part. 'Taking action' was introduced in the last chapter as 'the setting of the various knobs on the radio'. So we have already been introduced to the idea that there are a large number of possible knobs such as frequency, bandwidth, signal duration, modulation technique, power, etc. that can be set, but we have not looked at any details. In this chapter we look at the details and more explicitly at what actions are needed for the kinds of applications described in the opening chapter of this book.

To do this we need first of all to further build our knowledge about what actions are possible. A second important point of this chapter is to develop an understanding of the consequences of the actions taken. While we have stressed in the previous chapter that 'taking action' is not just about the physicality of the transmitted signal and can pertain to other aspects of the communication process such as higher-layer performance issues, management of battery lifetime of the node or the processing resources of the node, it is the physical interaction of the transmitted signal with other entities around it that is core to understanding the consequences of the actions that are taken. Essentially when a signal is transmitted it interferes with other signals and other signals interfere with it. Looking at consequences of actions is important for two reasons. Firstly, understanding consequence helps determine which actions will deliver the desired behaviour for the particular application of interest. Secondly, understanding consequence will also help understand the *undesirable* effects of actions. We begin the chapter by looking at the world in which actions take place and build an understanding of interference. We then move on to taking a brief tour of different communication systems to set the scene for what actions are possible. We then return to describing in detail the various actions that shape the behaviour of cognitive radios. Finally, we look at how settings for a cognitive radio (i.e. selection of actions) are communicated to the receiver.

3.2 Understanding the world in which actions take place

In an ideal world a transmitter will transmit a well-defined signal (whatever the waveform) that falls within a very specific range of frequencies and this crisp and clear signal will reach the receiver undistorted. The reality is far from this. Transmitted signals get distorted, corrupted, interfere unintentionally with other systems and cause all sorts of problems. The signals undergo distortion at the transmitter, en route from the transmitter to the receiver and at the receiver itself.

3.2.1 At the transmitter

Figure 3.1 shows (a) the ideal output of an RF transmitter, as well as (b) the real transmitted signal. The representations are somewhat simplified but the main issues are captured. The key point is that frequencies outside of the allotted frequency band are transmitted. These unwanted emissions are made up of *out-of-band emissions* and *spurious emissions*. Out-of-band emissions are emissions immediately outside the allotted bandwidth and are due to the modulation process. When information

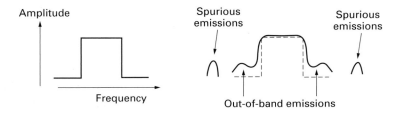

(a) Ideal output of the transmitter   (b) Real output of the transmitter

Fig. 3.1. The reality of RF transmissions.

is modulated onto the carrier for transmission, the modulation process causes the signal to be expand in frequency. Hence the modulated signal contains more frequencies than the unmodulated signal. Spurious emissions are all other kinds of emissions outside the allotted bandwidth. They can arise, for example, because signals can inadvertently couple together in the transmitter to produce undesired signals at new frequencies. A strong interfering signal, perhaps from another communication system, can be the cause of this coupling, which leads to the generation of signals at new frequencies.

Good filtering, before transmission, can remove some of the unwanted emissions. However, there is no such thing as a perfect filter and some unwanted emissions will always occur. Hence, from the very beginning of the communication process the signal can contain unwanted frequencies.

3.2.2 Along the way

Radio waves weaken as they travel away from the transmitter. In general, the higher the frequency, the faster the waves weaken. Determining how much they weaken over their journey can be challenging. The radio waves are affected by the earth itself, the atmosphere, the topography and natural and man-made objects such as foliage and buildings that are in the path of the waves. Again depending on the frequency of operation different effects come into play and different modes of propagation dominate. The spectrum is divided into groups of frequency bands, each given their own name as depicted in Table 3.1. Spectrum is typically discussed in these bands. The following description draws from [1], [2] and [3].

At frequencies in the 300 kHz to 3 MHz range, two basic modes of propagation, namely *ground waves* and *sky waves* come into play. Ground waves follow the curvature of the earth, and sky waves are reflected off the ionosphere. These modes of propagation can become more or less dominant on a temporal basis at some frequencies as well. Take AM broadcasting for example. The strength of signals from traditional AM broadcasters is likely to vary significantly from daytime to night time, from location to location (e.g. with latitude) and from season to season. In the daytime, coverage is provided by the ground wave and the service

Table 3.1. *Designation of frequency bands*

Frequency band	Frequency range
Extremely Low Frequency (ELF)	< 3 kHz
Very Low Frequency (VLF)	3–30 kHz
Low Frequency (LF)	30–300 kHz
Medium Frequency (MF)	300 kHz–3 MHz
High Frequency (HF)	3–30 MHz
Very High Frequency (VHF)	30–300 MHz
Ultra High Frequency (UHF)	300 MHz–3 GHz
Super High Frequency (SHF)	3–30 GHz
Extra High Frequency (EHF)	30–300 GHz

is comparatively reliable but relatively limited in range. In the night time, however, the radio signals in this range are carried beyond the horizon by reflections from the ionosphere. Because the ionosphere varies in height above the earth, the distances travelled by the waves can vary significantly.

The next highest range of the spectrum, the 3 MHz to 30 MHz range, is known as the shortwave region, and here, the ground wave component becomes less important and the signals are carried over vast distances by reflections from the ionosphere or even serial reflections between the earth and the ionosphere.

In the case of VHF (30 MHz to 300 MHz) and UHF (300 MHz to 3 GHz) wave propagation takes place mainly via direct and reflected ground waves. The VHF band has many important services such as VHF television, FM radio and a number of mobile services. The UHF band is a very desirable band for communications involving mobile and portable devices. VHF and UHF frequencies tend to be heavily influence by reflections and refractions of the signal that occur on its journey from transmitter to receiver. The signal experiences fading which can be both long-term and short-term in nature. Long-term variations in signal level are caused by shadowing effects due to energy-absorbing objects (i.e. buildings and other obstacles) along the propagation path between

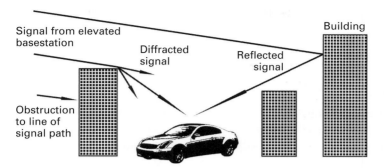

Fig. 3.2. Multipath and other effects.

the transmitter and the receiver. Shorter-term fading in signal level is generally caused by the movement of the transmitter or the receiver, or the movement of any obstacles in the propagation path. As the signal propagates it is reflected from various objects. A moving receiver, such as that depicted in Figure 3.2, will move through areas in which the reflections combine constructively to those in which they combine destructively and hence experience this rapid short-term fading. This effect is known as *multipath* and is a major challenge in mobile communications. *Rayleigh* and *Rician* fading can model the concept of multiple signals arriving at the transmitter. In a Rayleigh channel all signals arrive at the transmitter in equal strength whereas in a Rician channel one signal dominates.

The consequences of multipath can also be understood from a time perspective. As can be seen in Figure 3.2, signals can arrive at a receiver via very direct routes and via longer routes due to reflections that the signal experiences. This actually means that multiple copies of the signal, dispersed in time, can arrive at the receiver. The consequence of this is that if the next transmission takes place before these reflections 'die down', then interference can occur because there will be a jumble of the second transmission with reflections from the first transmission. This is known as *inter symbol interference*. The time needed to allow reflections to die down will depend very much on the environment the signal experiences 'along the way'. Some environments can have very short delay profiles while others have much longer.

Frequencies between 3 and 30 GHz are known as microwaves though the term microwave can also apply to frequencies above 1.5 GHz. These waves need line of sight for successful transmission. In essence this means that the transmitter and receiver must see each other. Any obstacles in the path would weaken the signal so much as to render it useless. The range between 30 and 200 GHz is referred to as millimetre wave propagation. Frequencies in this range are affected significantly by rain, snow and fog.

So different propagation mechanisms come into play at different frequencies, obstacles have varying types of effects and hence predicting the actual power of the received signal can be a challenging task. The field of propagation studies involves understanding the power losses that occur during the signal's journey, and many different equations exist to model the behaviour of the radio waves. There are simple methods for getting ballpark estimates of signal levels at the receiver as well as more complex path loss prediction models that can take the effects of shadowing and multipath, among other things, into account. In recent years, computer prediction models have become available which combine a path loss model with detailed mapping data, enabling the coverage of a transmitter in a particular location to be predicted. A more detailed discussion of these topics can be found in [1], but what is relevant for now is a broad understanding of the idea that the transmitted signal is affected on its journey, precisely how being frequency dependent. Irrespective of what exactly happens the signal that arrives at the receiver will be attenuated and can experience fluctuations in signal level.

3.2.3 At the receiver

The receiver will typically not just pick up the desired signal but also other signals that impinge on it but are not intended for it. The unwanted signals are often stronger than the wanted signals, which means that it can be very difficult to recover the information that has been transmitted. The incoming signal will also contain noise, and the receiver itself will also add noise to the signal.

If we look at this from a frequency perspective we can say that both unwanted out-of-band and unwanted in-band signals are present in the receiver. Out-of-band signals should in principle be removed at the receiver by an appropriate filter. However, just as in the case of the transmitter, completely ideal filters that just cover the allotted frequency band don't exist. This means that the receiver is capable of receiving signals from transmitters that operate in bands that are adjacent to the band of interest. This is known as *adjacent channel interference*. The exact level of adjacent signals that are received depends on the receiver design. A high-quality receiver will have a sharper filter and be able to reject more adjacent channel interference than a lower-quality one.

Unwanted in-band signals can also arise in a number of ways. First of all a transmitter operating at the same frequency of operation as the transmitter of interest can cause problems. This is called *co-channel interference*. Typically transmitters are spaced geographically so as to ensure that co-channel interference does not occur, i.e. the signal on the same frequency is so weak by the time that it reaches the receiver that it causes no problems. But as radio waves can be so unpredictable, at times co-channel interference does occur.[1]

Unwanted in-band signals can also arise from interfering signals that on the face of it appear not to be a problem as they seem far away from the band of interest. When certain signals at the receiver are very strong, a phenomenon known as *intermodulation distortion* can occur. This occurs when two signals suitably spaced in frequency combine in such a manner as to generate a signal at a third frequency that lands in the in-band. If the receiver is operating in its non-linear region of operation the intermodulation products can be very large and cause lots of problems by swamping the wanted signal. It is desirable that receivers operate in a linear fashion, i.e. that the output of the receiver is directly proportional to the input. However, real receivers have a range of operation which is

1 Readers may be familiar with the notion of a reuse distance in a cellular network. Reuse distances in GSM cellular networks specify the distance that must exist between any two transmitters operating at the same frequency, in order to avoid such problems. Co-channel interference is a big focus in GSM network planning precisely because frequencies are reused to get great capacities.

linear, and called a *dynamic range*, beyond which non-linear behaviour occurs. Dynamic range is dealt with in detail in Chapter 7. Really expensive receivers will be able to maintain linear behaviour over a wider range of power levels than cheaper ones.

On top of the unwanted out-of-band and unwanted in-band interferences that occur, a general desensitisation of the receiver can take place when any strong signals add to the overall noise floor at the receiver, even if not at the same frequency. In signal theory, the noise floor is the measure of the signal created from the sum of all the noise sources and unwanted signals within a measurement system. Adding to the noise floor means the received signal must be stronger in order to ensure detection. An interesting calculation of the effect of an increase in interference on a cellular system is made in [4] and is worth repeating here. Haslett shows that for the particular (very typical) cellular system described an increase of even a small amount in the effective noise floor of 1 dB (1 decibel: a relative measurement of power, on a logarithmic scale) due to interference will have a knock on effect of reducing the coverage range of each basestation. As pointed out earlier the noise floor has an impact on the lowest power at which a signal can be received. Increasing the noise floor means that low-strength signals that could once be received are now not possible to detect. In the particular example used, 30% more basestations are needed to give the same coverage that was available before the 1dB increase in noise floor took place.

Figure 3.3 attempts to put all these interferences in context. The diagram is inspired by the diagram in [5]. In the diagram a number of adjacent frequency bands in a given area are shown. A receiver operates in one of those bands and the outline in Figure 3.3(a) shows its receiver profile. Figure 3.3(b) shows how signals that are being transmitted by other systems, in the adjacent bands, can be picked up by the receiver. Figure 3.3(c) attempts to depict co-channel interference. The transmitter is not physically present in the geographical area but nonetheless the receiver can pick up its transmission. Figure 3.3(d) shows some intermodulation effects, due to the coupling of strong transmissions in other bands and Figure 3.3(e) captures the concept of desensitisation with the shaded area indicating a rise in noise floor below which signals cannot be received. The

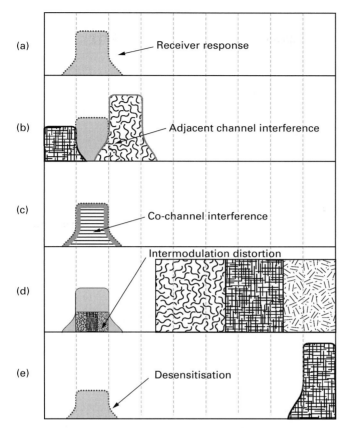

Fig. 3.3. A brief summary of interferences. *Almost depicts Prob(detection) vs. frequency*

interferences are summarised in Figure 3.4. While this may seem like overkill, interference is so important that it warrants the extra space. Figure 3.4 is from an Ofcom report on spectrum usage rights [6]. The figure captures the geographical aspects of the interference a little better than Figure 3.3.

3.2.4 Interference and cognitive radio

Obviously we need to understand interference so that radios, including cognitive radios, can be designed to deal with the effects of interference

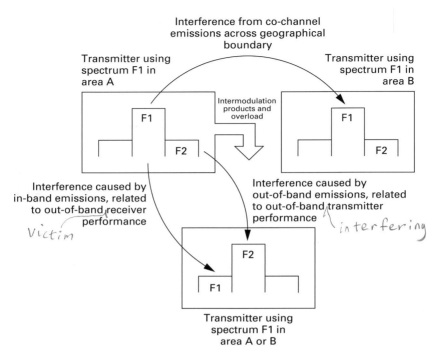

Fig. 3.4. The different interferences which must be taken into account.

and noise. But even more importantly we also need to understand the interference the cognitive radio might cause to other users. The main message here is that there is no action (or at least very few) without consequences for others and the consequence must be considered before action is taken. And a large number of the applications listed in Chapter 1 are about coexisting with other systems and hence very much about taking actions to mitigate against interference. In dynamic spectrum access applications, the cognitive radio must coexist with primary users or other dynamic spectrum access users. The success of cognitive radio usage of TV white spaces depends heavily on operation being on a *do no harm basis*. In Commons regimes different users must coexist with each other. In military scenarios, coalition forces must coexist with each other and this must happen quickly without the need for detailed spectrum planning. Likewise for public safety. In many of these application areas, the operating environment changes often necessitating a constant awareness of interference that feeds into choices of action. It is clear that *interference is much more of an issue in a cognitive radio world* as the ability to 'deal with it' is really what a large number of cognitive radio applications are concerned with.

3.3 A brief tour of communications systems

Communication systems are designed to be able to operate in the environments described in the previous section and to deal with interference as much as possible. Currently, when a system is designed, it is designed to operate in a given set of frequencies, use a particular waveform and share the wireless medium with other signals from fellow radios using a specific technique. While a description of the waveform transmitted and the manner in which the radio shares the wireless space with others by no means furnishes a complete picture of any given radio system, these two characteristics are very much defining characteristics of the system and hence are briefly discussed.

3.3.1 Waveform types

Communication systems can be classified under one of the following four headings: (1) narrowband, (2) spread spectrum, (3) multi-carrier or (4) ultra-wideband.

1. A narrowband communication system transmits and receives signals that, as the name implies, have a narrow bandwidth. A GSM (Global System for Mobile) network is an example of a narrowband system. In GSM the channels over which communication is established are 200 kHz wide. AM and FM systems are also narrowband. A narrowband system can be thought of as a single carrier system.
2. A spread spectrum communication system is one in which the transmitted signal is spread over a frequency band much wider than the minimum bandwidth needed to transmit the information being sent. Spread spectrum uses a number of different techniques such as spreading codes or hopping patterns to spread the normally narrowband information signal over the relatively wide band of frequencies. This

spreading of the signal decreases the potential interference to other receivers and at the same time increases the immunity of spread spectrum receivers to noise and interference. Some cellular networks such as 3G UMTS based systems used spread spectrum techniques.

3. A multi-carrier communication system is one in which the data to be transmitted is spilt into several components, and each of these components is transmitted over separate carrier signals. The individual carriers have narrow bandwidth, but the composite signal has a much larger bandwidth. The WiMAX standard is based on a multi-carrier communication technique.

Add:
Wideband
Comm.
System

4. An ultra-wideband (UWB) communication system has bandwidth exceeding the lesser of 500 MHz or 20% of the arithmetic centre frequency of the signal to be transmitted, according to the Federal Communications Commission (FCC). UWB transmissions are very low in power.

While all of these communication systems are important and will continue to be important, multi-carrier communication techniques feature heavily in much of the work that has been carried out in the cognitive radio world. Hence it is worth taking some time to describe a very popular multi-carrier communication system that is known as orthogonal frequency division multiplexing (OFDM).

Focus on OFDM

OFDM is a major multi-carrier transmission technique that is used in many wireless communication systems (e.g. Wireless LAN – 802.11, WiMAX – 802.16, digital audio/video broadcast systems such as Digital Video Broadcast – Handheld (DVB-H), Media FLO, etc.). In OFDM the information is transmitted using a large number of equally spaced subcarriers rather than one carrier alone. While each subcarrier transmits data at a low rate, the fact that there are a large number of subcarriers means that very high data rates can be achieved. The subcarriers have an important special property in that each subcarrier is orthogonal[2] to

---

2 To be orthogonal means to be at right angles to something. In the context of communication systems this means being completely distinguishable from something else. If orthogonality is lost the ability to distinguish two things from each other is lost.

every other subcarrier. This means the subcarriers can be very tightly spaced, making the system spectrally very efficient, while at the same time the orthogonality ensures that the receiver can separate the different subcarriers despite the tight spacing. In OFDM the data is modulated on to each subcarrier by varying the amplitude or phase, or both.

OFDM can be implemented in a very straight-forward manner using what are known as Fourier transform techniques. The Fourier transform provides a means of converting a signal from a time-based representation to a frequency-based representation; the inverse Fourier transform performs the reverse operation. Fourier transforms are used all the time in manipulating signals in radios. In essence the signal to be transmitted is prepared in the frequency domain, hence the parallel streams of data are manipulated separately and then converted to the time domain for transmission, with the reverse happening at the receiver. An OFDM system takes a data stream and converts it into a number of parallel streams corresponding to the number of subcarriers. Each stream is mapped to an individual subcarrier. The number of bits of the data stream mapped to each subcarrier depends on the modulation technique used. Following this, the inverse Fourier transform[3] is used to take this frequency-domain representation of the signal and convert it to the time-domain waveform that will be transmitted. At the receiver end, the time-domain signal is received and the Fourier transform is used to split the signal into its frequency representation so that the individual subcarriers can be demodulated and the data recovered.

OFDM has many advantages. As stated already, OFDM is spectrally efficient. It has an inherent robustness against narrowband interference. Any narrowband interference will affect at most a couple of subcarriers, and the information from the affected subcarriers can be recovered using error correction schemes. OFDM is also very good in multipath environments. This characteristic makes it very attractive for mobile communications, which experience much multipath. The transmitted OFDM waveform consists of a series of what are termed OFDM symbols. A

---

3 Note here that discrete versions of the Fourier transform and inverse Fourier transform are used.

technique is used that inserts a guard interval[4] between each OFDM symbol transmitted. The guard interval between each OFDM symbol allows for any reflections to die down before the next symbol is received making it easier to deal with multipath.

OFDM has disadvantages too, like any communication technique. One major difficulty is its large peak to average power ratio (PAPR). These large peaks cause saturation in power amplifiers, leading to intermodulation products among the subcarriers and disturbing out-of-band energy. Frequency synchronisation in OFDM systems is more complicated than that in single carrier systems. Carrier frequency offset disturbs subcarrier orthogonality and resulting inter-channel interference that severely degrades the demodulator performance.

There are other types of multicarrier systems: for example spread-spectrum-based multicarrier communication systems. These also have interesting properties and are useful in cognitive radios but the main ideas of multicarrier systems are covered in the OFDM discussion.

3.3.2 Multiple access techniques

The wireless medium is a shared resource. Hence when a group of users are using a given frequency band there needs to be some way of slicing up the cake so that each user gets a piece. The technical term used is *multiple access* – i.e. the means by which multiple users can share the common medium.

Some multiple access techniques sense if carriers are being used and attempt to avoid any collisions with frequencies that are in use. Carrier sense multiple access with collision avoidance (CSMA/CA) is one such example. Other techniques essentially 'draw a chalk circle' around each user's part of the spectrum resource. There are many different ways of doing this. The frequency band can be divided among the users and each user gets a certain slice each. This is known as frequency division

---

4 In reality a cyclic prefix is used – this is more than just a delay between symbols. It is like a guard interval that contains a repeated section of the OFDM, which is better able than an empty time interval to recover the received signal in the face of reflections.

multiple access (FDMA). Alternatively each user can access the band completely but for particular instances in time only. This is called time division multiple access (TDMA). Combinations of FDMA and TDMA exist in many systems such as GSM systems. Different frequencies are allotted to different cells and, within those frequencies, users get different time slots. The third main way is for each user to use the whole band at the same time but with a particular code per user. This is what happens in spread spectrum systems, the code being a spreading code or a frequency hopping pattern unique to each user. This is called code division multiple access (CDMA). Figure 3.5 captures the concept of code division multiple access in a unique manner.

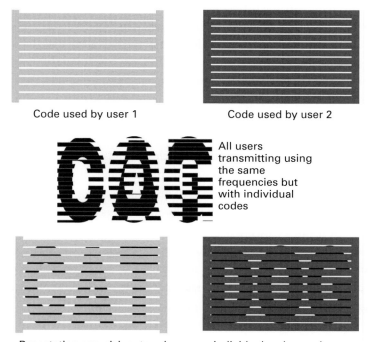

Code used by user 1   Code used by user 2

All users transmitting using the same frequencies but with individual codes

Basestation can pick out each user as individual codes are known

Fig. 3.5. Code division multiple access (CDMA) explained. The concept was designed by Goyal & Howard from Bell Labs and is used with their permission.

In multi-carrier systems different users can be accommodated by assigning them different subcarriers. It is possible to introduce a time division approach on any subcarrier as well. Multi-carrier CDMA systems also exist. Here we see mixtures of frequency and code mechanisms for providing multiple access. The exact details are beyond the scope of this book. Suffice to say the ways of slicing the cake are many. The only prerequisite for any slicing technique is that the different slices can be distinguished from each other.

3.4 The actions in detail

Typically radio systems are designed with specific applications in mind and for use at a predefined set of frequencies. The industry is very driven by standards mainly so that different manufacturers can produce systems that inter-operate with each other. A communications standard will typically specify that a radio needs to produce a certain waveform and share the wireless medium using a specified multiple access technique, among other things (such as frequency of operation, modulation technique, error coding, etc.). The standard limits the actions that can be taken by a radio. Cognitive radios can emerge within the context of standards. In this case a cognitive radio would produce the one waveform of interest and use a defined multiple access technique but could cleverly adapt its parameters or learn new behaviours within the bounds of the standard. And as we shall see later in this section there are very many options for adaptation, even based on a given waveform, and many more ways for radios to be cognitive even within some restrictions.

However, the real strength of a cognitive radio becomes clear when it can substantially change the shape of a waveform or use more than just one waveform and can more freely set its parameters of operation. An ideal cognitive radio might be capable of producing any type of signal of interest, narrowband, spread spectrum, multi-carrier or UWB in nature, and capable of exploiting whatever multiple access technique is needed.[5] It is conceivable that many of the military applications or the public safety

---

5 High levels of flexibility put high demands on the physical hardware in use at the RF frontend of the radio as well as on the digital aspects of the design.

applications detailed in Chapter 1 might demand such levels of complexity. Where inter-operability between different entities (such as different public safety sectors) is of major concern, the cognitive radio needs at least to be able to support the waveforms for the networks with which it wishes to inter-operate. But even commercial applications of cognitive radio call on a more open approach to the selection and shaping of the transmitted waveform and the more flexible use of available resources.

Table 3.2 maps the applications in Chapter 1 to a set of high-level actions. As the table shows, many of the actions, whether commercial, public safety or military based, centre on two activities:

1. The cognitive radio shapes its transmission profile and configures any other relevant radio parameters to make best use of the resources it has been given, while at the same time not impinging on the resources of others.
2. If and when those resources change, it reshapes its transmission profile and reconfigures any other relevant operating parameters.

The previous discussion on communication systems helps illuminate what is meant by resources. A cognitive radio will be given a set of frequencies or a set of codes or a set of times slots to use or some mixture of these. It will typically have these types of resources at its disposal over a particular spatial domain and for some time duration depending on the application at hand. For example, the time duration will be 'until the primary user returns' in the case of dynamic spectrum access regimes, or may be 'until the spectrum is traded' in the technology-neutral licence regimes or spectrum trading examples, or 'as long as the disaster lasts' for public safety interruptible spectrum regimes. The signal can be shaped to use these resources in its frequency content, its spatial footprint and its temporal profile. The available resources can be exploited by making sure the signal is created in such a way to make it robust for its journey as well as making optimal use of available capacity. And the available resources can be redistributed by giving other radios more time slots or frequencies or even by changing the method by which the division of resources occurs. Higher-layer actions can also help.

Table 3.2. *Some high-level actions for different applications*

Application area	High-level actions
Dynamic spectrum access	The cognitive radio identifies white space, shapes its transmitted signal to fit into the available space, making the most out of the white space. The shaping process is repeated as the white space changes.
Technology-neutral coexistence	The cognitive radio or cognitive network shapes its transmission profile to fit within its own spectrum assignment and to coexist with neighbours, making best use of its spectrum assignment, and reshaping should those neighbours change.
Spectrum trading	The cognitive radio shapes its transmitted signal to fit the spectrum resources obtained from the trading process, making best use of the acquired spectrum.
Low-maintenance self-forming networks	The cognitive radio or cognitive network shapes its transmission profile to fit with existing networks and to establish connectivity without manual intervention.
Military/public-safety self-forming networks	The cognitive radio or cognitive network shapes its transmission profile to coexist and inter-operate with others in line with whatever spectrum resources are available. It targets bandwidth or network time resources to where they are most needed and re-targets those resources as and when needed. It reshapes its transmission profile and re-targets resources as operations evolve.
Interruptible spectrum	The cognitive radio or cognitive network shapes its transmission profile to use available spectrum resources. It releases or grabs extra spectrum resources as and when needed, reconfiguring the network and reshaping the transmissions as appropriate.

Table 3.2. (*Cont.*)

Application area	High-level actions
Radio personalisation	The cognitive radio shapes its transmission profile and manages the communication process in line with user preferences and available spectrum and network resources. As user preferences change, reshaping of the transmission profile and reconfiguring of how the network is used takes place.
Delay tolerant networking	The cognitive radio or cognitive network shapes its transmission profile and manages the communication process in general to match applications to changing spectrum resources.
Preferred network connectivity (e.g. for better tariffs)	The cognitive radio shapes its transmission profile to be capable of communicating in the network of interest
'Green Radio'	The cognitive radio shapes its transmission profile and manages the general communication process to use resources, be they battery resources, spectrum resources, network resources or any other, in a manner that is most energy efficient.

3.4.1 Taking action from a frequency perspective

Actions relating to the manipulation of a transmitted signal from a frequency perspective tend to involve, in general, making sure the transmitted signal contains only the frequencies it should and making sure the receiver gets rid of unwanted frequencies. To aid the discussion it is perhaps helpful to think about shaping the frequency content of a signal in the context of some applications. Consider the idealised spectrum shown in Figure 3.6 and the possible options open to a radio that wishes to communicate in the indicated white spaces (marked A and B). The radio can be one that is dynamically accessing white space. Alternatively the radio can be a self-configuring basestation that wants to coexist with existing systems that are using the marked frequencies.

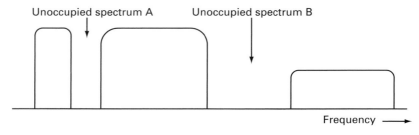

Fig. 3.6. Some spectrum with identified unoccupied parts (a simplified and somewhat stylised representation).

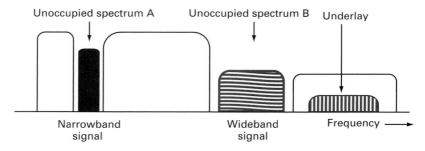

Fig. 3.7. Options for filling the unoccupied space – an ideal cognitive radio could produce all signal types.

We begin by looking at the options for the general shape of the signal that is to be transmitted. Figure 3.7 is based on Figure 3.6 and shows how these various signal types could be used. A narrowband signal could be used to communicate in white space A, while a wideband signal could be used in white space B – this could be a spread spectrum. An underlay using either some kind of spread spectrum technique or a UWB system could be used in the occupied spectrum regions.

It was pointed out earlier that a cognitive radio may not necessarily be able to produce all waveforms. This is where multi-carrier techniques prove useful. What makes OFDM particularly interesting in this context is the ability to manipulate the frequency content of the transmitted waveform. Subcarriers can be dynamically turned on and off to provide a means of coexisting with other systems. So, for example, if we return to Figure 3.6, a cognitive radio could transmit an OFDM signal with 64

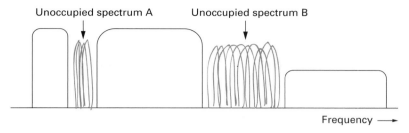

Fig. 3.8. A varying number of OFDM subcarriers can be used to suit available spectrum.

subcarriers if it is using white space B while a signal with very few subcarriers could be transmitted if the cognitive radio can only use white space A. This concept is shown in Figure 3.8. Note that we have assumed in the diagrams thus far that each white space is used by a different secondary user or different basestation. It is possible also for one cognitive radio to simultaneously transmit in both white spaces by creating an OFDM signal that consists of subcarriers that are nulled in the centre of the waveform, in the position where the spectrum user exists (between white space A and white space B). This is known as non-contiguous OFDM and is also of great interest in the cognitive radio world. Though there are challenges relating to maintaining orthogonality etc., it is an interesting approach. Using OFDM as a good way of controlling the frequency content of a signal also has disadvantages. As mentioned previously, OFDM has what is known as a high peak to average power ratio. This means that at a receiver there is a higher amount of incident power than in the case of spread spectrum communications. Hence the possibility of driving the receiver into its non-linear region of operation and generating very damaging intermodulation effects is increased.

A cognitive radio has both analogue and digital parts. The production of the waveform of interest, be it any of the forms listed above, tends to happen in the digital part of the radio. The data is typically converted from a stream of ones and zeros to the waveform of interest and then converted to analogue format for transmission. The preparation of the waveform is sometimes referred to as baseband processing as the signal is manipulated at baseband and it is only in the analogue part of the radio that it is up-converted to the frequencies that are used for transmission. To be able to convert to any one of a range of frequencies, frequency-agile analogue hardware is needed. This has implications for the antenna used (typically antennas cover a set range of frequencies), for the filters in the analogue frontend of the radio and for the other RF components. For example, the filters need to be tunable so as to move to the frequency bands of interest. The antenna needs to cover all bands of interest or alternatively be tunable and reconfigurable itself. The wider the band of frequencies needed to be covered, the more demands are put on the hardware. Chapter 7 will take up these points again when discussing the current state of the art in this field.

As has been discussed already, the act of transmitting a signal can cause interference to others. To avoid interference, not only must the signal be centred around the desired frequency for transmission and the bandwidth fixed so as to remain in-band, the out-of-band emissions must also be controlled. In Figure 3.9 the users of the white spaces are causing some adjacent channel interference to the primary users. User 1 and user 2 are both causing adjacent channel interference. This can cause problems for the existing users as explained earlier in the chapter. Obviously, as in the transmitted signals in Figure 3.9, the OFDM signals can cause adjacent channel interference to the primary users. In reality, they do not conform to the very neat depictions shown in Figure 3.8.

Action can be taken both in the digital and analogue parts of the radio to remedy this. In the analogue part of the radio, the power could be reduced, to bring down the level of the main transmitted in-band signal and, as a

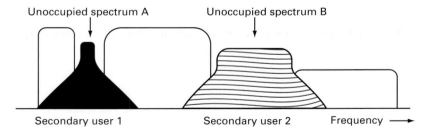

Fig. 3.9. The cognitive radios can cause interference to their neighbours.

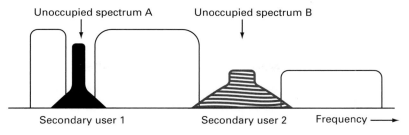

Fig. 3.10. User 2 uses a narrower range of frequencies, to avoid interfering with primary users. User 1 uses better filter techniques.

result, the level of the out-of-band emissions. The linearity of the components in the analogue frontend and the quality of the filters have a great bearing on the output, and the level of quality will be fixed. However, the presence of cheaper filters with less good performance characteristics does not mean that a cognitive radio cannot fit into the white space. It just means that a more cautious use of the available white space must take place. So, for example, to remedy the situation, users 1 and 2 in Figure 3.9 may be forced to use a much narrower band of frequencies than is available, resulting in the situation depicted in Figure 3.10. So, while some adjacent channel interference still occurs, it has been greatly reduced.

On the digital side a whole range of actions can be taken. Consider OFDM again. It is possible to null more subcarriers near the edge of the frequency band to cut down on interference. This can be thought of as a self-induced guard band. It may be that quite a large number of subcarriers must be nulled but this will depend on the quality of the filters in the cognitive radio hardware, and the quality of the RF frontend will have an impact on the number of carriers that must be nulled.

The blunt removal of subcarriers to shape the frequency profile of the transmitted signal is not the only possibility. It is also possible to apply digital filtering to shape the signal to have a much better profile. Hence a whole range of filtering actions are possible. Other choices of operating parameters in the digital domain also have an impact on the frequency profile of the signal. For example the modulation technique in use has a bearing on the transmitted signal from a frequency perspective. However, it is also possible to filter a signal before it is modulated to reduce the bandwidth it takes up. Filtering allows the transmitted bandwidth to be significantly reduced without losing the content of the digital data. This improves the spectral efficiency of the signal. There are many different varieties of filtering. The most common are raised cosine, square-root raised cosine and Gaussian filters. Sometimes filtering is desired at both the transmitter and receiver. Filtering in the transmitter reduces the adjacent-channel-power radiation of the transmitter, and thus its potential for interfering with other transmitters. Filtering at the receiver reduces the effects of broadband noise and also interference from other transmitters in nearby channels. Obviously filtering adds to the complexity of the transmitter and the receiver.

There are a large number of knobs that can be set that will have an effect on the frequency content of a transmitted signal. The general waveform of the transmitted signal can be selected and parameters of the waveform can be manipulated. Even just using OFDM as an example we see that the number of subcarriers it wishes to use (which actually corresponds to the act of setting the Fourier transform size), the modulation scheme it uses to map the information to each subcarrier, whether all subcarriers are active or not (either to create more spectrally limited signals or to create non-contiguous patterns of interest), how the subcarriers are windowed, etc. can all be dynamically set. The filters on the RF frontend can also be actively controlled. The power levels can be set.

One point worth nothing at the end of the first discussion on actions is that actions or knobs identified here are discussed in terms of their impact on the frequency content of a signal. The reason for choosing the setting of any one knob though need not be on the basis of one effect it has. For example, the choice of modulation scheme can also be driven by channel throughput. What drives the choice of setting or how a balance is achieved is covered in Chapter 5.

3.4.2 Taking action from a spatial perspective

To elaborate on how the spatial footprint of a transmitter and receiver may be manipulated the example of the use of white space will also serve to highlight the issues of interest. In Figure 3.7 and in the previous section, the transmitted signals are described in frequency terms. However, as we know, a propagated signal also has a spatial dimension. Figure 3.11 looks at the transmitted signals from the viewpoint of existing users and cognitive users in terms of signal range. The existing user could be a TV broadcaster and the cognitive user could be making use of TV white space for example. Or the existing user could be licensed spectrum user and the cognitive user could be a self-configuring system trying to coex- ist with it. As in the frequency representation, simplified diagrams are used but they nonetheless capture the main issues. In Figure 3.11(a) the existing basestation is using frequency band F1 to communicate with two users. The circle is a simplified depiction of the range of the base- station. The two existing users are within range of the transmitter and all is well.

The problem arises when a cognitive basestation that is outside the range (i.e. in a different geographical area) of the existing transmit- ter wishes to communicate, as in Figure 3.11(b). The cognitive base- station is not aware of any transmissions on F1 and decides to use this

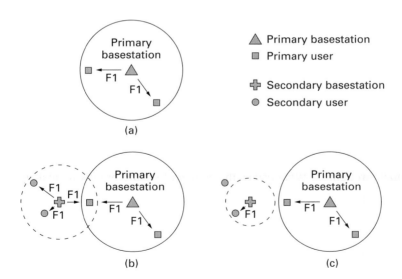

Fig. 3.11. A spatial view of interference.

frequency band when communicating with its two users in Figure 3.11(b). Because of this, one of the existing receivers experiences interference from the cognitive user. In other words co-channel interference occurs. One solution is for the cognitive basestation to change frequency of operation. But we will leave this to one side for the moment. The cognitive basestation can reduce the power of its transmissions, to avoid interfering, but as a result the coverage footprint reduces. However the penalty is that the more distant cognitive user now gets no coverage, as shown in Figure 3.11(c). If we recall Figure 3.4 the two boundaries, namely the spatial and frequency boundary of the signal, need to considered. And what we are doing here is shaping the spatial footprint, and not just the frequency content as discussed in Section 3.4.1.

Transmit power control can be useful in other scenarios as well. Suppose now that the two cognitive transceivers in Figure 3.11 are communicating with their cognitive basestation on a frequency band other than F1 and hence not interfering with the existing user, transmit power control may need to be used here too. This is to make sure cognitive users are not actually interfering with each other. Consider the case in which some kind of spread spectrum communication is ongoing. In this case, as all users communicate simultaneously but using different codes, it is possible to envisage scenarios in which one user drowns out the other. All signals should arrive at the basestation's receiver with the same signal power, but if both cognitive users were to transmit using fixed power levels, the user closest to the basestation would dominate and faraway users could not get their signals heard in the basestation. This phenomenon is called the *near-far affect*. Hence we see that transmit power control is not just to do no harm to other systems but also to avoid cognitive users harming each other.

The coverage patterns in Figure 3.11 are based on omnidirectional antennas being used on all radios.[6] The antenna also offers great

---

6 A plot of field strength vs. direction is called the antenna's radiation pattern. An omnidirectional antenna has a radiation pattern that is the same strength in all directions. In general, antennas generate stronger electromagnetic waves in some directions than others, and more specific directions. A directional antenna, as mentioned in this section, is one that is specifically designed to target a given direction.

possibilities for shaping the footprint of the transmitted signal and for getting the receiver to be more receptive of particular signals. A directional antenna is one designed to have a gain in one direction and losses in others. Hence directional antennas allow the energy in the signal to be directed towards a target of interest and, more importantly, away from other targets. A directional antenna that focuses its energy in the direction of the two secondary receivers and away from the primary receivers in Figure 3.11 is an alternative solution that can ensure both users get coverage. A variety of techniques exist for dynamically steering the beam of the directional antenna. This can be useful in mobile scenarios.

It is also possible to create more sophisticated antenna beam patterns that contain multiple strong lobes and multiple nulls. This is known as beamforming and it requires the use of an array of antennas rather than a single antenna. Each antenna element is fed a separate signal whose amplitude and phase are weighted to produce the pattern of interest. This can be done in a dynamic adaptive fashion (typically in digital beamforming the manipulation of the amplitude and phase of the signal fed to the array elements actually happens in the digital domain). Adaptive beamforming systems for communications are sometimes referred to as *smart antenna* systems. Cognitive radios using adaptive digital beamforming have an enormous amount of scope for tailoring the signal to be transmitted.

Figure 3.12 depicts some sample *antenna radiation patterns*. Figure 3.12(a) shows an omnidirectional antenna, whose signal is strong and equal in all directions. In Figure 3.12(b) the beam of the antenna

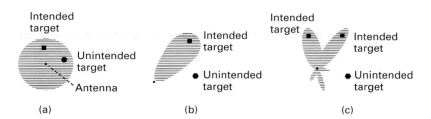

(a)   (b)   (c)

Fig. 3.12. Different antenna radiation patterns that can be created depending on the type of antenna in use.

is very directional so it can be pointed towards a target of interest and avoid an unintended target, in this case the primary user. Figure 3.12(c) shows a more sophisticated antenna pattern that is a result of beamforming. The beams are directed at two different secondary users and away from primary users.

The directionality or the beamforming can be put to good use at the receiver as well as the transmitter. In the case of the receiver the antenna pattern will encourage good reception of signals for desirable directions and place nulls in directions of strong interferers. The polarisation of the antenna can also play a role in terms of spatial footprint. Certain polarisations will give greater penetration of the signal for a given transmit power.

The knobs for controlling the spatial footprint of the transmitted and received signals are knobs for setting power levels and any knobs related to controlling the radiation pattern of the antenna. There are fewer knobs than those for manipulating frequency content but they are no less important.

3.4.3 Taking action from a signal robustness perspective

As mentioned at the beginning of Section 3.4, there is a balancing act to be carried out between making sure a transmitted signal is received while at the same time avoiding causing meaningful interference to other systems. The last two sections on the frequency content of the transmitted and received signals and their spatial footprints very much focused on interference mitigation. If we now look more towards ensuring that a transmitted signal survives the journey to the receiver, there is another range of actions which can be taken by the cognitive radio. Again these actions can take place in the digital part or the analogue part of the radio.

The choice of waveform was discussed in the context of taking actions from a frequency perspective. Making a system robust could also drive this choice. For example OFDM systems do well in multipath environments. Once the waveform is chosen, the signal can be prepared for transmission in a manner that makes it better able to withstand the difficulties it experiences on its journey. One such mechanism for doing this is error coding. Noise and interference cause errors in the signal, as we have already learned. Error-correction codes allow errors to be corrected. In general error coding schemes involve adding extra bits to the bit stream in such a manner as to make it possible for the receiver to spot when errors have taken place and to correct the errors as well. There are very many different coding schemes, each with different capabilities and implementation complexities. If the channel is very noisy, more robust and complex schemes are needed, which tend to require the addition of more bits than the less robust schemes. Like any schemes, coding schemes have their limits and can only correct a finite number of errors. It is possible to design cognitive radio transceivers that allow the dynamic setting of the parameters of the coding system and hence a range of actions can be taken in this respect.

The information to be transmitted can be further manipulated, depending on the underlying transmission technique in use. For example, the data can be scrambled to make it more robust to fading. Scrambling involves rearranging the order in which the data is sent. If a series of transmitted packets are corrupted, for example, the corruption will manifest itself in small pockets called bursts throughout the received data rather than in one big pocket. The former scenario lends itself to corrective action while the latter does not. It is also possible to take preemptive action to compensate for the noise that the receiver will add through the use of a technique known as pre-distortion. In the context of OFDM communication systems, there are also choices that can be made in relation to the length of the cyclic prefix and the number of pilot tones needed (some subcarriers can be used as pilot tones to help the receiver lock on to the signal), among other things, which have a bearing on how robust the signal will be to the interference it experiences. Coded OFDM can also be used. Coded in this case refers to the fact that error coding is added, as described in the previous paragraph.

The antenna can also play a role in making a signal more robust through the use of what are known as *diversity techniques*. The basic idea is to help the receiver by giving it the opportunity to get a strong signal. This is done by finding different ways of transmitting and receiving the same signal. Suppose an individual wants to relay a long and complicated verbal message to another individual by means of a messenger. If one messenger is used, there is a chance that the complete message may be forgotten or jumbled. If instead a number of different messengers are used there are two options. The first is that there is an increased chance that one of the messengers will get the message correct. The second option allows for the combining of all the partially correct messages received, to reconstruct the full message. The fact that the sender of the message diversified the sending method and the receiver avails of the diversity increases the chances of a successful transmission.

The diversity process involves finding some physically differentiable means of sending the same signal to the receiver together with advanced signal processing techniques. Space diversity, in which a single data stream is replicated and transmitted over multiple antennas that are separated in space, is one example of a diversity technique. The term 'multiple input, multiple output' (MIMO) is used to describe an antenna system that has multiple antennas at the transmitter and the receiver. MIMO systems can be used to achieve this kind of diversity. In a space diversity approach, the redundant data streams are each encoded using special mathematical algorithms that ensure each transmitted signal is orthogonal to the others. This means that the chances of the individual streams self-interfering can be reduced and the capability of the receiver to distinguish between the multiple signals can be improved. With the multiple transmissions of the coded data stream, there is increased opportunity for the receiver to identify a strong signal that is less adversely affected by the physical path. Alternatively, the receiver can use combining techniques to combine the multiple signals for more robust reception. In the context of the discussion on taking action, a cognitive radio can select which diversity technique to use (provided the physical infrastructure such as multiple antennas is in place) as well as select or fine-tune the signal processing algorithms which are used in the transmission and the reception of the diverse signals. The singling out of the strongest signal or the combining of the signals in a useful manner happen in the receiver.

On the face of it, many of the actions listed here are actions that are taken in various communication systems that already exist. It is perhaps worth emphasising again that, in the case of the cognitive radio, many of the actions will be taken in a much more dynamic sense and in response to the prevailing conditions rather than in a pre-determined manner. In sum, making the signal robust for transmission and achieving robust reception involves playing with a whole host of radio characteristics at both the transmitter and the receiver in the digital and analogue domains of the radio.

3.4.4 Taking action from a data throughput perspective

We now look at the actions that can be taken to maximise the throughput of the communication session. A variety of actions can be taken in this context.

For example, source coding can be used to compress the data that is sent. There are choices to be made here in terms of the exact actions taken. Highly compressed data will mean that the data can be sent to its destination more quickly than uncompressed data. However, if the channel conditions are poor and a deep fade occurs in the transmitted signal, much more data will be lost during that fade than in data that is uncompressed.

The modulation scheme in use also has a large impact on the throughput of the system. The signal bandwidth for the communications channel needed depends on the symbol rate. Symbol rate is the bit rate divided by the number of bits transmitted with each symbol. So, for example, a BPSK modulated signal uses one bit per symbol while a QPSK modulated signal uses 2 bits per symbol. If more bits can be sent with each symbol, then the same amount of data can be sent more quickly. However, the higher number of bits per symbol is only possible in good signal conditions. Typically the aim will be to achieve a particular BER (bit error rate) and the modulation scheme that can deliver that under the prevailing conditions is the modulation scheme of choice. It is worth noting that the choice of modulation scheme is not just about maximising capacity but, as mentioned already, has a bearing on the frequency content of the signal. Choice of modulation scheme is a very good example of an action that impacts on the transmission of a signal in many different ways.

Capacity is also obviously affected by the actions taken to make the signal robust. So while data can be compressed to increase the through-put of the signal or while the best possible modulation scheme can be used, error coding or other such processing of the transmitted signal will decrease the data throughput. Again all of these actions need to be looked at individually as well as collectively in terms of achieving the particular goal of the cognitive radio.

It is also possible to make use of antennas to increase the capacity of the system. The MIMO systems mentioned earlier were used to make sure that the transmitted signals were more robust to the propagation environment through the exploitation of diversity techniques. MIMO systems can also be used for spatial multiplexing in order to increase capacity. In spatial multiplexing, a high-rate signal is split into multiple lower-rate streams and each stream is transmitted from a different transmit antenna in the same frequency channel. If these signals arrive at the receiver antenna array with sufficiently different spatial signatures, the receiver can separate these streams, creating parallel channels for free. Spatial multiplexing is a very powerful technique for increasing channel capacity when the signal-to-noise ratio is good. The maximum number of spatial streams is limited by the lesser in the number of antennas at the transmit-ter or receiver. It is possible to make use of MIMO systems that avail of spatial multiplexing when the signal-to-noise ratio is good enough to do so and to use the diversity techniques mentioned in Section 3.4.3 when the signal-to-noise ratio is poor. A cognitive radio should be capable of taking the appropriate action and in that sense is an ideal entity to really make use of a MIMO system. MIMO systems are not currently in very widespread use, but cognitive radios could lead to wider take-up.

In terms of maximising throughput, we again see that a number of actions are open to the cognitive radio. The prevailing conditions deter-mine what actions are suitable and some actions can be contradictory to each other. It is important to choose wisely.

3.4.5 Taking action from a resource distribution perspective

Looking back at Table 3.2 we see that many applications involve targeting resources dynamically where needed. The actions in this section simply fall into two different categories. Thee first category involves setting the knobs of the individual radio to make use of, for example, more time slots or more frequencies or more codes.

The second category is the actions relating to changing how the divisions are made. This boils down to changing the multiple access technique in use. The multiple access issues are handled in the MAC layer, the layer above the PHY layer. From the outset of Section 3.4, the resources given to the cognitive radio were described as if a static MAC were in place. In particular it was stated that 'A cognitive radio will be given a set of frequencies, a set of codes or a set of times slots to use, or some mixture of these.' It is possible to take an action that involves changing the kind of MAC in use or actually changing features of the particular MAC in use dynamically. In fact the concept of an adaptive MAC features heavily in the world of cognitive radio. We will return to this in the next chapter as we need to know more about sensing to expand this point.

3.4.6 Taking action from a higher-layer perspective

To date all that has been discussed is mainly at the level of the physical creation and manipulation of the waveforms to be transmitted at the PHY layer. Aspects of MAC layer manipulation have also been discussed because it is at the MAC layer that coding and error correction occur and that multiple access issues are handled. As stressed in previous chapters, radios are not standalone entities but are devices that operate in networks. To be part of a network the radio node needs other functionality. Hence the term *cognitive node* can often be more apt than cognitive radio. With a node perspective, taking action can therefore include a much richer range of actions than those associated with the physical layer of the radio as it is possible to think of ways in which a cognitive radio (or node) can set parameters associated with the higher layers of the system.

Beyond the PHY and MAC layer there are many other layers of the system. The networking layer sits on the MAC layer which sits on the physical layer. The routing of data through the network is handled by the networking layer. Above the networking layer are other layers related to the applications that run on the radio. There can also be layers that handle security and other functionality. The options for taking action at these various layers are too numerous to list and would also necessitate a more detailed description of each layer. But, as brief examples, stronger or lighter security measures can be selected or different routing protocols can be used (for example, the node can switch to an ad hoc mode from an infrastructure-based mode when necessary). The ability to more dynamically control routing, or adaptive control routing, can be very important in public safety applications, if links go down. Or indeed in the context of Green Radio where the various route and routing options are given energy audits as a means of choosing the best options. The possibilities for finding new and more knobs to be set is indeed large.

3.4.7 Taking action together

A final perspective from which we can analyse the actions of a radio is in the context of any one cognitive radio having limited scope for action. This seems like a contradiction, but as stated at the outset of Section 3.4, not all cognitive radios will be able to take all the actions described thus far. Any given cognitive radio will have a certain number of resources available to it and have a certain set of capabilities. There are two options for a cognitive radio when it comes to having limited resources and capabilities. In the first instance it can make the best use of whatever resources and capabilities it has. So, for example, a cognitive radio that does not have a MIMO antenna system will make the best use of its single antenna. The second option open to a cognitive radio with limited resources is to take some kind of collective action with others that can be considered as *supportive* radios. In this context the cognitive radio can be considered to be a distributed entity. So, for example, a cognitive radio could make use of a nearby radio with a MIMO system. It would, over a very short distance, transmit to the more capable radio, which could in turn relay the signal over longer distances using the diversity capabilities of the MIMO system. The limited cognitive radio in this case can be seen as having some kind of parasitic relationship with the more advanced system. Even beyond this it is possible to create what is termed a *distributed MIMO system* and use a few single antenna systems to act in unison to benefit from MIMO advantages.

Obviously the use of multiple physical radios to form one virtual radio is challenging and, given that cognitive radios have still to manifest themselves in reality, it seems that this notion is very distant. However, the cognitive paradigm does allow us to rethink what a radio is and hence it is worthwhile mentioning the notion of a distributed or virtual radio, if only briefly.

3.5 Communicating the transmitter configuration details to the receiver

In discussing the actions taken by a cognitive radio, it is clear that many of the actions involve reconfiguring the transmitter in a way that necessitates a corresponding reconfiguration on the receiver side. A major challenge for any radio that reconfigures is how to relate the details of the reconfiguration to the receiver. This challenge is even greater on initialisation of communications because, for many cognitive radio applications, the frequency of operation is a widely varying parameter. We start at this point of initial communication and then move to the broader issues associated with communicating reconfiguration details to the receiver.

Frequency rendezvous is the term used to define the process in which two radios 'meet' on a common frequency in order to communicate. For traditional radio systems this is not a problem. For example when I want my FM radio to rendezvous with RTE Radio 1, the main Irish radio station, I simple set the receiver to the frequency at which I know the radio station operates. I can do this because I have *a priori* knowledge of what that frequency is.

A GSM network provides a better example of how rendezvous occurs in more complex networks. In a GSM network a mobile device must be assigned a channel from a range of available channels in order to communicate. However, while the mobile device will know in advance the broad range of frequencies in which the channel lies, it will not know which channel to use. To rendezvous with the network a GSM mobile device needs to associate with a particular basestation and synchronise in time with the basestation so that it can obtain a channel for communication purposes. To deal with this challenge GSM systems do not use all available channels for data and voice traffic. Instead a certain number of channels are set aside and dedicated for use as *control channels*. Only control information is transmitted on these channels while the remaining (and majority of) channels are used for voice and data traffic. This kind of control channel is sometimes called an *out-of-band control channel* as the control information uses completely separate channels from the voice and data traffic. The GSM mobile device scans through all the control channels, the frequencies of which it knows *a priori*, to seek out the information it needs to appropriately configure itself, synchronise and establish communication.

Control channels known as *in-band control channels* also exist. In in-band control channels, as the name suggests, the data traffic and the control information use the same channel. A WiMAX system uses this approach. The control information is placed on the channel in such a way as to ensure it can easily be identified as control information and not confused with data. Here again the frequencies of operation are known *a priori* and WiMAX users know where to look for the control information they need. The WiMAX devices can get all the information they need to appropriately configure themselves to enter the network.

As is clear from the discussion thus far, control channels of one sort or another are common in many communication systems. It is therefore logical to suggest that a control channel be used in cognitive communication systems both as a means of establishing initial communication (frequency rendezvous) as well as for communicating subsequent changes in configuration. A control channel on one level does not sit so easily with the underlying philosophy of cognitive networks. The best way of making this point is by asking the question 'what is a channel?' In old analogue cellular networks the control channel was a dedicated frequency. In a GSM network it is a slightly more complicated entity. In GSM, a 25 MHz frequency band is divided, using a FDMA scheme, into 124 carrier frequencies with a 200 kHz spacing. Each carrier is time-divided using a TDMA scheme. A TDMA frame has eight time slots. A channel maps to the recurrence of a time slot every frame. In a CDMA network a channel is a code. The point of all this is simply that a channel must be defined in the first instance for the channel to be used. This presupposes the use of particular technologies, which is a bit at odds with the more open philosophy of the cognitive radio using whatever communication techniques best suit prevailing circumstances. Having said that, it does not significantly limit the functioning of a cognitive network to define a control channel that could be used on start-up to 'bootstrap' communications. A simple out-of-band channel may be very useful in this context and subsequent communication could perhaps use in-band signalling for further control information.

Control channels, however simply defined, are more easily dealt with in certain cognitive radio applications than others. Any application that simply extends more traditional networks to include cognitive features could easily use control channels. Spectrum trading is an example where a control channel would be very appropriate. However, in applications that involve dynamic spectrum access techniques, the use of a control channel is very problematic. A good application to illustrate the difficulties is secondary unlicensed cognitive radios using unoccupied primary licensed spectrum. Here secondary users access spectrum opportunistically. A cognitive mobile device wishing to rendezvous with another device will not know in advance what frequency that device is using. The use of a control channel which could provide information about who is using what frequencies seems a logical solution. However, the concept of setting aside a dedicated control channel does not easily fit in with the notion of dynamic spectrum access. Different spectrum is available at different times and in different locations. There may be no 'universal' free set of frequencies that can be set aside to act as a control channel. There are some suggestions for tackling this issue. There is talk of regulators actively allocating spectrum as a cognitive control channel. However, in a very open system in which all cognitive users operate over different frequency ranges, more than one control channel may be necessary. There have been other suggestions for control channels that include using separate, easily available radio technologies to provide a control channel such as ISM band-based technologies (e.g. 802.11). There are without doubt downsides to these approaches. A universal control channel or set of control channels may easily become congested. The use of a separate and fixed radio technology to access a control channel may have cost issues. There are also questions about the 'control of the control channel'!. However, a static control channel or set of control channels is still a possibility. It would more than likely be necessary to reduce the degrees of freedom of the cognitive network to cope with such an entity, but not necessarily to a severe extent.

Many of the community interested in cognitive radio tend to react against the idea of a static control channel. Most of the suggestions for alternative approaches are still in the research realm. There is some research, for example, that has focused on the use of dynamic control channels for dynamic spectrum access networks. In this approach the aim is not to find spectrum that can universally be used as a control channel but instead can be used locally. Most other ideas, though, put the use of a control channel to one side and look for other means of performing frequency rendezvous and other techniques for communicating reconfiguration details to the receiver. However, as will be seen in the following paragraphs, any suggestions, just as in the case of a static control channel, involve making one kind of assumption or other about the radio capabilities.

The main rendezvous ideas that do not make use of a control channel involve some kind of scanning or searching through the available spectrum, by the radios wishing to communicate, to find a suitable channel. To understand how this happens we look at an analogy. Consider two people who want to have a conversation but who do not speak each other's language. Hence they need to call on translators. Suppose now they each have 10 translators and each translator is capable of translating the original language of interest into one other language only. Person A and person B select a translator each. The translators attempt to establish communication. However, no communication is established until the translators are translating the original languages into one common language. If person A and B are lucky they happen upon the correct combination quickly. If they are not lucky they have to try every combination – (Person A translator A1, person B translator B1), (person A translator A1, person B translator B2), . . ., (person A translator A1, person B, translator B10), . . . , (person A, translator A10, person B translator B10) – until they find a solution. This is akin to what happens when two radios, radio A and radio B, attempt to rendezvous without a control channel. Radio A works its way through a sequence of possible transmitting frequencies and radio B through a sequence of possible receiving frequencies. If both radios happen upon the same frequency at the same time then rendezvous is established. The translator example serves to emphasise that radio B must be listening on the frequency that radio A is transmitting for communication to be established as well as to emphasise the fact that a number of different combinations may need to be sampled before rendezvous can happen.

Even without a control channel there are still restrictions in play. The definition of a channel must be set. There must be some kind of rough synchronisation between the radios. Radio A must be trying frequency A1 when radio B is listening on that frequency. The transmission and reception do not need to coincide exactly in time but an overlap must exist. While restrictions exist there are promising signs and progress has been made [7] in finding clever sequences for searching that achieve rendezvous more quickly than using the more random approach described in the previous paragraph.

Other techniques exist, again in the early research stage, that use some kind of 'physical signalling' to aid the rendezvous process. These techniques insert some kind of digital watermark or digital signature into the transmitted signal that identifies the transmitter and also reveals something about the frequency at which the signal is operating. Using the terminology used earlier, radio A would insert such a watermark into its transmissions and Radio B would look for this watermark. The digital watermark or signature is designed to allow radio B to find it, if it exists, very efficiently. Details of these techniques are beyond the scope of the book but Sutton *et al.* [8] provide a good description of this approach. Here again restrictions are in place. The solutions by Sutton *et al.* are based on the use of OFDM as the signatures are generated by manipulating OFDM subcarriers. And there is also an assumption that each user has some kind of unique identity that has somehow been centrally administered.

Once rendezvous has taken place, by whatever mechanism, and communication is established the radios can then reconfigure themselves in a more sophisticated manner. If a control channel exists, configuration information can be exchanged over the channel. Alternatively, some kind of in-band control signalling can be agreed upon between the different parties and used for further exchange of configuration information. Another possibility is to use techniques that support the automatic detection of configuration changes. For example, there are a number of solutions for automatic modulation detection. This means that a transmitter can change modulation technique as desired and the receiver will automatically follow suit.

Undoubtedly there are open challenges when it comes to frequency rendezvous and to generally communicating configuration changes from node to node. The message, however, is that there are possibilities for using control channels and possibilities that involve alternatives to control channels. Any and all of these possibilities dictate certain aspects of the communication process at the initial bootstrapping phase. It seems sensible that a focus on standards in cognitive radio would focus on these aspects of the process while allowing the wider cognitive communication processes to be as open as possible.

3.6 Conclusions

As stated at the outset of the chapter, taking action for a cognitive radio means setting its transmit and receive parameters to produce a desired transmit and receive behaviour in order to achieve a desired goal or set of goals. We have seen that a key issue for all actions is to balance the desire for communication with any unwanted interferences that may arise as a result of that communication, and we have gained some insight into the complexity of the interplay between different radio parameters. We have seen that these are opportunities for action at the physical and higher layers of the cognitive radio node. The exact implementation of a cognitive radio will determine which actions are open to it. Not all cognitive radios may facilitate adaptivity beyond the physical layer. Not all cognitive radios may avail of MIMO antenna systems and use diversity or spatial multiplexing techniques. Even bearing this in mind, however, the range of parameters that can be manipulated can be quite significant.

The burning question that arises from this discussion is how the cognitive radio determines the values for the operating parameters? How does it know what spectrum to use, what power levels to set, which modulation or coding schemes are best, which resources are limited, what its own capabilities are in the first place, etc.? The first part of this answer is that the cognitive radio needs to gain an understanding of the prevailing conditions and the environment in which it is operating to be in a position to make decisions as to what must be done. In other words the cognitive radio must make and collect observations of the world around it and process those observations to get the awareness it needs to take the correct and appropriate actions. This is the theme of the next chapter.

References

1. D. Parsons, *The Mobile Radio Propagation Channel*. London: Pentech Press, 1992.
2. T. S. Rappaport, *Wireless Communications: Principles and Practice*, Prentice Hall Communications Engineering and Emerging Technologies Series. Upper Saddle River, NJ: Prentice Hall, 1996.
3. P. J. Weiser and D. N. Hadfield, Spectrum policy reform and the next frontier of property rights, *George Mason Law Review*, **60**:3 (2008), 549–609.
4. C. Haslett, *Essentials of Radio Wave Propagation*, Cambridge Wireless Essentials Series, Cambridge University Press, 2007.
5. V. R. Petty, R. Rajbanshi, D. Datla *et al.*, Feasibility of dynamic spectrum access in underutilized television bands, in *2nd IEEE International Symposium on New Frontiers in Dynamic Spectrum Access Networks, 2007*. Dublin, 17–20 April 2007, pp. 331–9.
6. Ofcom, Spectrum usage rights: Technology and usage neutral access to the radio spectrum. Office of Communication, 2006. Available at: http://www.ofcom.org.uk/consult/condocs/sur/ (accessed October 2008).
7. L. Da Silva and I. Guerreiro, Sequence-based rendezvous for dynamic spectrum access, in *3rd IEEE International Symposium on New*

Frontiers in Dynamic Spectrum Access Networks, 2008. Chicago 14–17 October 2008.

8. P. D. Sutton, K. E. Nolan and L. E. Doyle, Cyclostationary signatures in practical cognitive radio applications, *IEEE Journal on Selected Areas in Communications (JSAC)*, **26**:1 (2008), 13–24.

4   Observing the outside world

4.1 Introduction

To study observations, we return yet again to the definition of the cognitive radio laid out in Chapter 1 and note once more that 'A cognitive radio is a device which has four broad inputs, namely, an understanding of the environment in which it operates, an understanding of the communication requirements of the user(s), an understanding of the regulatory policies which apply to it and an understanding of its own capabilities.' Getting these four inputs is what we mean by the phrase 'observing the outside world'.

We can further detail some of the observations that are needed if we go through the various action categories outlined in the last chapter. To take action from a frequency perspective the cognitive radio must observe which signals are currently being transmitted, which channels are free, the bandwidth of those channels and perhaps whether the available channels are likely to be short lived or more durable. To take action from a spatial perspective, the cognitive radio needs to make observations about the spatial distribution of systems that must be avoided, or the spatial distribution of interferers and of the target radios. The cognitive radio needs to be able to monitor its power output and the power output of other systems. To take action to make a signal more robust or to maximise the throughput of the transmitted signal, the cognitive radio needs to make observations about the signal-to-noise ratio (SNR) at the target receivers, about the bit error rates and about the propagation conditions experienced by the transmitted signal (e.g. delay spread, doppler spread). To take actions that focus on maximising the capacity of a channel the cognitive radio needs to make observations about the channel. Observations about the cognitive radio's location, the date and day, whether it is indoors or outdoors, may feed into any of the above actions. Observations about the kind of network in which it is operating (dense, sparse, mobile, stationary, infrastructure-based, ad hoc) might inform higher-layer actions. To take actions at higher layers, the cognitive radio needs to be able to note routing details, cache details, traffic flow patterns or congestion patterns. The essential point is that there are numerous observations of relevance.

When it comes to the 'how' of making observations, in simple terms a cognitive radio either makes the observations itself or gets the observations from other entities. To make its *own autonomous local observations* the cognitive radio can use whatever software and hardware it has at its disposal – we shall see later that different aspects of the software and hardware come into play when gathering observations. Broadly speaking there are four general approaches here:

1. Firstly the cognitive radio can get observational data as a natural by-product of its normal mode of operation. This is the simplest scenario. So, for example, time and date are typically available in many operating systems. The transmit power setting may be readily observable. As a more complicated example a node in a network may know how many neighbours it has as a consequence of some aspect of its routing protocol and hence node degree is readily available. In other words the observations in this category are simply available as part of the normal operating process of the radio and no elaborate processing is needed to access the observations.[1]

2. Secondly a cognitive radio can make observations via dedicated extra hardware such as a GPS, simple sensors (temperature, pressure, acceleration) or battery meters. A GPS, for example, is capable of determining location (to a certain degree of accuracy) which in turn can be used to determine velocity over time. A temperature sensor could provide information on operating conditions and, together with the output from a battery meter, indicating remaining operating lifetime, could for example be used to manage radio resources. There is no assumption that all cognitive radios will have such extra equipment, but it is certainly a possibility that some will.

---

1 However, a mechanism must exist within the radio framework to make these observations widely available to whatever parts of the cognitive radio subsequently need to process them.

3. Thirdly a cognitive radio can use specialised signal analysis techniques for extracting relevant observations from signals that the cognitive radio captures, either through the radio with which it normally communicates or through additional radio sensing equipment. The analysis techniques could be used to determine the presence or absence of signals, to discover details of the features of the signals which are present, to determine interference levels, or to evaluate signal-to-noise ratios, etc. In other words, the analysis would generate a set of observations that all tend to be connected with understanding the RF environment or with what Haykin [1] would term *radio scene analysis*.

4. Fourthly a cognitive radio can learn from experience about the world in which it is operating. It can learn where network coverage is poor. It can learn the preferences of the user of the system. It can learn the behavioural pattern of primary users.

To get observations from 'other entities', a cognitive radio can, as an obvious first choice, get information from another cognitive radio. Cognitive radios can share and distribute information among each other leading to the cooperative production of collective observations and a more global view of what is going on. As we shall learn later on in the chapter, making collective observations can be crucial to the functioning of the cognitive radio. Alternatively, the cognitive radio can avail of some external service, such as a centrally controlled database, to glean information about which system is operating where and on what frequencies etc. Though the dependency on external dedicated infrastructure may not be desirable, for some applications the only sure way of getting valid observations may be from an external service.

As these various approaches indicate, some observations are straightforward and simple (e.g. reading a GPS meter, noting a transmit power, reading an external database), while others involve much more complex approaches that involve heavy-duty signal processing and perhaps even collaboration between different nodes (e.g. determining a white space).

All the four different categories of input to the cognitive radio are important. However, in terms of gaining an understanding of the environment in which the cognitive radio operates, observations about which other systems are currently operating are the most fundamental. So whether the cognitive radio is a secondary user looking for a white space, or a public safety radio that wishes to coexist with another public safety system, or a cellular basestation wishing to pool extra spectrum, knowledge of what spectrum is occupied or free is essential. We refer to the process of determining how the spectrum is occupied as *spectrum sensing*. Spectrum sensing is a very challenging process as in many cases the signals that must be detected have very low powers and the noise levels at which they must be detected are high. Very often the observe stage of the cognitive cycle is synonymous with the spectrum sensing task. The spectrum sensing task involves determining what spectrum is occupied in frequency as well as in geographical (spatial footprint) terms and in doing so details of the propagation conditions also become known. With this in mind the bulk of the chapter is devoted to spectrum sensing.

Spectrum sensing is tackled on an individual radio basis as well as a collaborative basis in the discussion that follows. The issues relating to collaboration can be generalised to other collaborative actions. At the end of the chapter we return to looking at observations that are obtained from outside entities.

4.2 The spectrum sensing challenge

Spectrum sensing in very broad terms involves the detection, by a given receiver, of the presence of a transmitted signal of interest. The ability of a cognitive radio to dynamically access white spaces that dynamically appear is predicated upon its ability to detect these white spaces in the first place. The ability of a network of cognitive radios to coexist with other existing networks is predicated upon being able to sense the existence of the other networks. Whether they be two technology-neutral service providers, two public safety or two military networks, the detection of the presence of the other system for the purpose of altering the transmission characteristics of both systems to allow them both to exist side-by-side without interfering with each other is called for. Furthermore, not only must the cognitive radio sense spectrum occupancy levels, continuous monitoring of the spectrum may subsequently be necessary. In the case of secondary users occupying white space, they must be on the lookout for the return of the primary user. Coexisting systems may have to adapt to changes in neighbouring systems and this necessitates 'keeping an eye out' as well.

While cognitive radios are not simply radios for dynamic spectrum access, as stated on numerous occasions already, we will nonetheless focus on the case of a secondary user wishing to use unoccupied spectrum left vacant by a primary user. This scenario captures many of the key issues that arise in spectrum sensing. The primary user, as you will recall from Chapter 1, under one example regime, is the licensed user of the spectrum and has priority access to the spectrum. Secondary users, i.e. the unlicensed users, can only occupy the spectrum when unused by the primary user.

4.2.1 Sensing accurately

The first objective of any spectrum sensing technique is to *accurately* detect the presence of existing transmitters – in this case the primary user. The literature in the field tends to express the problem of postulating whether a primary user is present or not as a *hypothesis* test. The null hypothesis states that there is no signal in a certain spectrum band, i.e. there is just noise. On the other hand, the alternative hypothesis states there is more than noise there and the signal includes transmissions from licensed users.

The problem is therefore all about determining whether the null or alternative hypothesis is true, i.e. to actually detect the presence of the primary user. If there is in fact no primary transmitter present and the cognitive radio observes that there is, this is known as a *false alarm*. If on the other hand there is a transmitter present and the cognitive radio does not observe its presence then this is known as a *missed detection*.

The aim of any observation mechanism used to detect the presence of a primary transmitter is to make sure that the number of false alarms and the number of missed detections are as low as they possibly can be and hence the detection rate is as high as possible. A false alarm can lead to a missed opportunity. For example, if the cognitive radio thinks spectrum is occupied when in fact it is not, then the secondary user will not transmit.

While this is undesirable, a missed detection has even more serious consequences. A cognitive radio, thinking spectrum is free, may transmit its own signal and in doing so greatly interfere with the undetected primary user. For cognitive radios to become accepted and to gain widespread use, the ability to accurately sense white space is very important.

Typically the false alarms and missed detections are expressed as probabilities. Different spectrum sensing techniques may achieve different probabilities of false alarms or missed detections. Therefore one way of comparing techniques is to use these metrics. The key here is the ability of the technique used to cope with undesired signals and noise.

4.2.2 Sensing over the appropriate range

The second objective of any sensing technique is to be valid and usable *over the appropriate detection range*. To a certain extent this is an extension of the accuracy requirement expressed in terms of the range from a primary transmitter at which it must be detected.

Any transmitter has what is known as a range of decodability. The range of decodability is the maximum distance at which the receiver can properly receive and decode a transmitted signal. The decodability range depends on the power of the transmitter as well as the sensitivity of the receiver. Typically when a primary user is guaranteed service from a primary transmitter, that guarantee will be based on the fact the primary user has receiving equipment of a defined level of sensitivity. The secondary user must be able to detect the presence of the primary transmitter over the decodability range. In other words it cannot be less sensitive than the primary user receiver.

4.2.3 Sensing in a timely fashion

The third objective of any sensing technique is to sense the existence of the primary user in a *timely fashion*. Figure 4.1 demonstrates this aspect of the problem.

In Figure 4.1, one particular channel is depicted. The primary user's occupancy of the channel is shown on top and the secondary user's

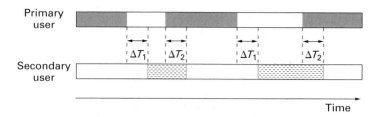

Fig. 4.1. Timing issues for spectrum sensing. The secondary user must get in promptly and get out even more promptly on the return of the primary user.

occupancy of that same channel is shown on the bottom. When the primary user vacates the channel a secondary user occupies it, and the secondary user in turn vacates the channel on return of the primary user. The time ΔT_1 is the time taken for the secondary user to observe that the channel is actually free and to take action and use it. The time ΔT_2 is the time taken for the secondary user to observe that the primary user is back and to subsequently vacate.

ΔT_1 and ΔT_2 are important metrics. If the observation process is very long and hence there is a large ΔT_1, then this can lead to very inefficient use of white space. In some cases the opportunity to use the space may have passed by the time that opportunity has been noted. If ΔT_2 is very large, the amount of interference caused to the primary user may be unacceptable. Note that a bound can be set on the metric ΔT_2 which can be used as a part of the specification of the maximum amount of interference permitted on the primary user by the cognitive radio. The complexity of the spectrum sensing algorithm will have a bearing on these timing metrics.

We see therefore that the challenge is not just about accurately determining if spectrum is free, but doing it in a timely manner.

4.2.4 Meeting the objectives in the face of interference

The objectives of accuracy, timeliness and sensitivity are made particularly difficult to meet when the reality of the communication process is taken into account. From the discussion of interference in Chapter 3 we know that the bottom line is that a receiver picks up wanted and unwanted signals. The unwanted signals consist of noise and interference and can drown out and mask the wanted signals as well as wholly disrupt the functioning of the receiver. To detect the presence of a primary user, sense has to be made of all this mess. The secondary user has to be able to distinguish whether just noise and interference is present or whether in fact a primary user signal has been picked up. The secondary user has to do this with a high degree of accuracy and in a timely fashion. The secondary user has to be sensitive enough to detect the required level of primary user activity. The more mess that exists, the more difficult and the more time-consuming it can be to analyse the incoming signal.

4.3 The basic sensing system

With the challenges of the spectrum sensing task in mind, we now move on to look at the sensing process in more detail. Figure 4.2 shows the main blocks of a basic spectrum sensing system. The system consists of a wideband RF receiver that captures the incoming signal. The receiver must be capable of capturing the entire range of frequencies in which the secondary user plans to operate (needs appropriate RF components and a suitable wideband antenna). The captured signal is then converted from analogue to digital and then processed in the digital domain. A variety of

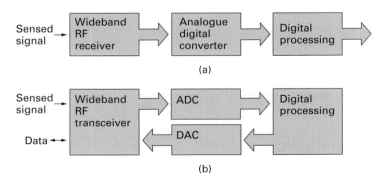

Fig. 4.2. The basic building blocks for spectrum sensing. (a) Dedicated sensing, (b) Integrated sensing.

signal processing techniques can be used to examine the content of the signal and determine the presence of spectrum users. Once white space has been identified the cognitive radio can transmit in that white space.

The cognitive radio can contain dedicated sensing circuitry, in which case a sensing system such as that shown in Figure 4.2(a) would be a separate entity from the main radio that is used for communication purposes. The other option is for there to be a single radio only. In this case the radio must divide its time between the sensing and communications functions. Figure 4.2(b) captures this approach.

There are advantages and disadvantages to either approach. Dedicated sensing hardware can add cost to the system as it imposes additional complexity and power constraints on the radio. Interferences can occur between the sensing hardware and the communications hardware. On the other hand a system that is used for both sensing and data communications will not be able to sense continuously. It may be necessary to turn off communication transmissions even with dedicated sensing hardware to get accurate readings of spectrum occupancy.

The quality of hardware in use has a bearing on the level of signal which can be sensed. A low-noise wideband frontend will mean a better signal is received than in the case of a receiver that is less well made. The ability of the receiver to remain in its linear range of operation is a good example of where the quality of the receiver matters. As was explained in Chapter 3, the receiver will be subject to additional effects of intermodulation, if the energy received is sufficiently strong to drive the frontend components into non-linear regions. A received signal that has experienced much distortion due to intermodulation effects may falsely appear to have no white space at all. Highly linear frontends are desirable but are more expensive hardware. In the case of a frontend architecture that uses separate hardware for sensing, the sensing receiver may need to be of better quality than the receiver used for communications purposes.

The analogue-to-digital conversion will also have an impact. Analogue-to-digital converters (ADCs) have two primary characteristics, sampling rate and dynamic range. Sampling rate is the number of times per second that the ADC measures the analogue signal and will determine how big a bandwidth can be digitised. One of the major limitations of a radio frontend's ability to detect small signals is its dynamic range. Dynamic range describes the range of the input signal levels that can be reliably measured simultaneously, in particular the ability to accurately measure small signals in the presence of the large signals. The dynamic range is a function of the number of bits in the ADC's digital output and the design of the converter. For example, an 8-bit converter can represent at most 256 signal levels, while a 16-bit converter represents up to 65 536 levels. Generally speaking, device physics and cost impose trade-offs between the sample rate and dynamic range. The dynamic range of the spectrum sensing system should be as high as possible.[2] In Chapter 7 the hardware that can be used in cognitive radios will be discussed further.

Once the cognitive radio captures a received signal and converts it to the digital domain, a number of signal processing techniques can be brought to bear to analyse the content of the received signal. The techniques are described in the next section. Essentially the idea is to measure some aspect of the captured signal that gives a clue as to whether the primary user is there or not. Bearing in mind the previous discussion about interference, the trick is to be able to do this even when the captured signal is quite distorted.

While a very good quality RF frontend will help get a cleaner signal and make life easier, it is really the signal processing technique which is at the centre of the spectrum sensing process. It is possible to have techniques that are robust to interference and therefore relax the constraints on the RF frontend. This underlines a lot of what cognitive radio is about, i.e. the ability to make the best use of whatever the situation the radio finds itself in, through as clever an approach to manipulating and processing the incoming signals as possible.

There are three main signal processing techniques that can be used on the captured signals in order to detect the presence of a primary user: energy detection, matched filter detection and feature detection. Most

2 To give an example, to receive distant signals in the presence of strong nearby signals, a cellular basestation receiver must have wide dynamic range. For example, GSM specifications call for receivers that can accurately digitise signals from -13 dBm to -104 dBm in the presence of many other signals, which is a -91 dB dynamic range!

techniques tend to fall into one of these categories, even if they are not labelled as one or other of these. In order to choose an appropriate technique, an understanding of how sensitive each technique is to unknown noise and interference, and an understanding of the implementation complexity associated with the technique, help.

4.3.1 Energy detection

The simplest way of analysing a received signal for the presence of primary users is to use an *energy detector*. As the name implies, it senses the amount of energy in the signal received by the cognitive radio. If the signal received has more energy than a certain threshold level of energy, then the received signal is deemed to contain transmissions from primary users. The simplicity of the energy detector makes it a very attractive option.

One technique used to do energy detection is based on the use of the fast Fourier transform, which translates a signal from a time domain to a frequency domain representation. It can be thought of as a means of determining the power in each frequency of the signal resulting in what is known as the *power spectral density* of the received signal. This is essentially a plot of energy versus frequency for the range of frequencies contained in the received signal. If the power level is deemed over a threshold value at a given frequency, then the presence of a primary user at that frequency is assumed.

To get an accurate evaluation of the energy in the signal at the frequencies of interest, it is necessary to get an average value of the received signal. The longer the time over which the signal is averaged, the more accurate the answer. Quite a number of samples are generally needed to get good results. This has implications for the ΔT_1 and the ΔT_2 parameters mentioned in Figure 4.1.

The energy detector has a number of drawbacks. The threshold level can be difficult to set correctly as the threshold depends on the noise levels which can vary considerably. One solution would be to employ an adaptive threshold level but this can be difficult too. Or the cognitive radio could learn the best threshold levels over time. The energy detector will just give an indication that there is energy in a given band – so, for example, in the case of dynamic spectrum access, it won't be clear whether that activity is due to a primary user or to some other source of interference. An energy detector also does not work for spread spectrum signals. In a spread spectrum communication system, the transmitted signal is spread over a frequency band much wider than the minimum bandwidth needed to transmit the information being sent. There are different kinds of spread spectrum signals. Code division multiple access (CDMA) is an example of a spread spectrum technique that is used in cellular networks – this was mentioned in the previous chapter. In the case of spread spectrum signals the power levels are very low and the signals tend to be not easily distinguishable from the noise. It is likely that in the case of spread spectrum signals an energy detector would have lots of missed detections.

The energy detector, however, does remain attractive because of its simplicity, and therefore work is being done on designing more robust energy detectors. Suggestions have been made for using the guard bands to estimate the noise/interference in the primary band, and gain robustness to interference uncertainty.

4.3.2 Matched filter

The energy detector is used when details of the primary user are not known and the idea is to perform a general check for activity levels. If details of the signals that are to be identified are known, then coherent techniques can be used, as opposed to the non-coherent energy detector technique. The terms coherent and non-coherent refer to the need for the receiver to synchronise itself (or not in the non-coherent case) in frequency and phase with the transmitter. Coherency is needed to demodulate the received signal for analysis of its content. In communication systems the transmitted signals are modulated, i.e. some characteristic of the carrier wave is made to vary in accordance with an information-bearing signal wave (the modulating wave). Demodulation is the process by which the original signal is recovered from the wave produced by modulation. Achieving coherency requires extra complexity in the receiver.

A matched filter is obtained by correlating a known signal, or template, with an unknown signal to detect the presence of the template in the unknown signal. In this case the unknown signal is the signal received by the cognitive radio. The known signal is the signal of the primary user. This means that the cognitive radio must know in advance what kind of primary user signals it wishes to detect. The cognitive radio in this case would store details of the primary user signals of interest in memory and compare these details with the received signal in an attempt to detect the presence of the primary user.

The process of comparing the received signal with the known signal or, more correctly stated, the process of correlating the received signal with the known template involves, as mentioned already, the demodulation of the received signal. The stored details of the primary user signal can help with the demodulation process. However, a lot of effort is needed to achieve coherency with the primary user signal, in order to actually carry out the demodulation. There are many techniques that can be used to achieve coherency that make use of, for example, pilots, preambles, synchronisation words or spreading codes in the primary signal. So, while the need for coherency has implications for the complexity of the receiver, coherent detection is possible.

An advantage of the matched filter, as it is based on coherent detection, is that it can quite quickly reach a decision as to whether a primary user is present or not. Speed of detection is typically measured in terms of the number of samples of the incoming signal that must be obtained before the presence or absence of a primary user can be determined. In the order of $(1/SNR)$ samples are needed in the case of the matched filter.

A matched filter is considered to be an optimal solution as the matched filter optimises the signal-to-noise ratio. The major disadvantage of the matched filter approach is that a dedicated receiver is needed for every type of primary user signal that may need to be detected. If the application domain is quite restricted the matched filter approach may be feasible as there may only be a small number of primary signal types to detect. In general, however, this disadvantage renders the matched filter not that useful for spectrum sensing.

4.3.3 Feature detectors

An alternative approach involves the use of a *feature detector*. A feature detector works on the basis of knowing something about the signal that is being detected and using some kind of suitable feature of that signal to aid in the detection process. TV signals have specific kinds of physical features that only TV signals have. Mobile phone signals have other features that are unique to them. Hence features detectors are signal processing algorithms that focus on identifying features that can be easily recognised and used to prove a signal exists.

One well-known feature detector is a *cyclostationary feature detector*. So what is a cyclostationary feature? We first start with the idea of a *stationary* signal. In mathematical terms we say that a stationary process is a random process whose statistics do not vary over time. The 'statistics' of the signal in question here are things like the mean value or the variance of the signal. So a stationary signal has a mean value that pretty much stays the same over time. A cyclostationary signal on the other hand has statistical properties that vary in a cyclical manner. So in this case the mean value of the signal may have some kind of periodic behaviour and therefore you would see a different mean depending on the period over which you calculated that mean.

The reality is that when you manipulate a signal to prepare it for transmission you introduce cyclostationary behaviour into the signal. Modulation is a good example. When a carrier wave is modulated the resulting signal exhibits statistical properties that vary cyclically. The cyclical rate or cyclic frequency (as it is more formally called and which is often denoted as α) at which the statistical properties vary depend on the exact modulation technique used. The upshot of this is that if I observe a particular kind of cyclical behaviour in the statistics of the received signal then I can deduce that the signal was modulated in a particular manner. This not only confirms that a signal is actually present but also, for example, gives me the opportunity to say that this is a mobile phone signal as distinct from a wireless LAN signal, as I will expect a certain kind of modulation in one that is not present in the other. Pulse trains, hopping sequences, cyclic prefixes and other features also cause the statistics of a signal to vary cyclically, i.e. cause the signal to exhibit cyclostationarity.

The next question is how do I observe the cyclostationary characteristics of a signal? Though we have been speaking about the time-varying statistics of a signal so far, such as the mean and variance, it turns out that it is also possible to do an analysis in the frequency domain as well as in the time domain.[3] The examination of the signal in the frequency domain allows us to spot the periodicities that have been introduced into the signal because of the cyclical behaviour of the statistics. In mathematical terms an entity known as a spectral correlation function (SCF) is generated to help spot this. The SCF is a two-dimensional function that shows the strength of every frequency component in a signal for varying values of α, the cyclic frequency. So if a feature such as a given modulation type has been used on the signal, then the cyclical variations introduced in statistics of the signal should manifest themselves as a frequency component at the cyclic frequency of interest. For interest it is worth noting that the SCF at an $\alpha = 0$ is actually the equivalent of what would be generated by the energy detector. Hence the cyclostationary feature detector gives more information than the energy detector.

The cyclostationary feature detector has two main advantages. Firstly it is very good for detecting low-powered signals in bad signal conditions. Spread spectrum signals are very low powered and are in common use. The cyclostationary feature detector really shows its value in the detection of these signals for very bad signal-to-noise ratios, where the energy detector will find it very difficult to tell the difference between the noise and the signal. The cyclostationary feature detector can tell the difference, as noise does not typically contain cyclostationary features. Secondly the cyclostationary feature detector, as already pointed out, deduces more about the received signal. For example different kinds of modulated signals (such as BPSK, QPSK, etc.) have very similar power spectral density functions. This means an energy detector would not be able to tell the difference between a primary user that uses BPSK and

3 Recall from earlier comments that a signal can be explored in either of the domains and that the Fourier transform is used to go from the time to the frequency domain. Both representations are equivalent.

one that uses QPSK.[4] However the spectral correlation functions of a BPSK and a QPSK modulated signal are different. Thus not only can the presence of a primary user be detected, features of the primary user signal can be identified. Bear in mind that, if features are to be identified, the cognitive radio needs to know what features to look out for.

While the cyclostationary feature detector has its strengths, it is not without disadvantages. It is computationally complex and requires significantly long observation time. This means that many samples of the received signal have to be averaged together to get a good SCF. Hence the value of the timing parameters ΔT_1 and ΔT_2, in Figure 4.1, may be so large that the technique can not be used in some application domains. Obviously care can be taken to implement the technique in as optimal a fashion as possible. For example, very efficient implementations are now possible on multi-core systems (multi-core platforms are discussed in Chapter 7). There can also be problems if strong interfering signals in adjacent bands or other effects such as non-linearities make the SCF appear to have the features you are looking for when in fact it does not.

4.3.4 The process in its entirety

As stated at the outset of the description of these three approaches, most sensing techniques are based on one or other of the three mechanisms mentioned here though the two most popular approaches tend to be the energy detection and cyclostationary feature detection. The spectrum sensing process need not just use one mechanism. For example, energy detection can be used to get a broad sense of the business of a band and can be followed by a deeper feature analysis of more targeted frequency ranges of interest. It is likely that a real cognitive radio will have a suite of techniques at its disposal and not only be able to choose the most appropriate technique for the task at hand but also be able to choose suitable parameters for operation on the chosen technique (e.g. threshold levels for the energy detector).

---

4 Depending on the application it may not be necessary to determine anything more than the presence of the primary user.

A cognitive radio can also learn the behavioural patterns of the primary user and use the information to predict occupancy. In the next chapter learning is one of the topics discussed. Spectrum occupancy patterns are a very good example of a behaviour worth learning.

4.4 Standalone or non-cooperative spectrum sensing

The simplest approach to spectrum sensing is for the individual node to sense what spectrum is available using any of the techniques described above. Suppose, for example, that a secondary user basestation is looking for free spectrum. The basestation alone would determine what spectrum is free and use that to provide service to its secondary users. This is known as non-cooperative sensing. Non-cooperative sensing takes place when individual radios, acting locally and autonomously, carry out their own spectrum occupancy measurements and analysis. Therefore each cognitive radio determines whether or not the primary user is present.

4.4.1 Non-cooperative sensing example

An example of a system using non-cooperative sensing is the DARPA XG radio. The DARPA XG research programme began in 2000. Part of the programme involved designing a radio that can coexist with legacy military radios. The resulting DARPA XG radio has the capability to detect the presence of legacy military devices and to avoid interfering with the legacy users by transmitting on unoccupied frequencies. The DARPA XG node therefore avails itself of dynamic spectrum access. The programme is an extensive research programme and many aspects of the work are relevant to the topics we cover in this book. For the purposes of this chapter we focus on the DARPA XG sensing capabilities. The details in the following paragraphs are taken from Seelig [2].

The DARPA XG node has dedicated sensing hardware and software that is based on an energy detection approach. A low-noise programmable high-bandwidth detector provides an estimate of power spectral density over 100 MHz with frequency resolution of down to 12.5 kHz. The power spectral density is obtained by getting the FFT of the received signal. Rapid internal signal processing allows an 8192 bin × 14 bit FFT to be computed in 1000 microseconds. The detector's range is 30–2500 MHz, samples at 18 GHz/s; has a user selectable resolution bandwidth of 12.5, 25, 50, 100 or 200 kHz. In the case of the DARPA XG project, the detector is typically configured for use in the VHF and UHF ranges. A discone antenna is used with the device as this can cover the required range of frequencies. A discone antenna is usually mounted vertically, with the disc at the top and the cone beneath. It is omnidirectional, vertically polarised and exhibits unity gain. It is a very wideband antenna and can cover a frequency range of 25–1500 MHz. Hence it adequately covers the bands under test. The DAPRA XG node can detect a weak signal in the presence of strong interferers. The details of this and of the legacy users can be found in [2].

4.4.2 Limits on non-cooperative sensing

The DARPA XG node is a good example of the successful implementation of an energy sensing technique in a non-cooperative manner. However, there are physical limits to the ability of an individual radio to detect the presence of a primary user. Another way of expressing this is that there will typically be a signal-to-noise ratio beyond which it won't be possible to use a specific technique (for a given accuracy). The term *SNR wall* is sometimes used to capture this notion. As we learned already, more advanced techniques can be applied but all these do is extend the limit rather than remove the limit completely. Even in some cases where it is possible to detect a signal, the detection process can take so much time as to render the process useless. To understand this it is important to realise that the various detection processes need to average the received signal over a number of samples to get a good idea of what is really out there. If large numbers of samples need to be averaged then the detection process can be very long (and this is even before taking into account the complexity of the detection process itself). Of all the limits that exist though, the *hidden node problem* is one of the most serious.

The hidden node problem

The power of a transmitted signal decreases with distance from the transmitter. Hence if the signal travels a long distance then it will be very low in power. If a cognitive radio is unable to detect the transmitter for this reason then there is no problem. The cognitive radio is far enough away from the transmitter to reuse the transmitter frequencies. However, because radio signals can suffer from fading on the journey from the transmitter to the receiver it is possible for a cognitive radio to be near a transmitted signal and not detect its presence. Fading was already discussed in Chapter 3. Recall that a radio signal experiences long-term and short-term fading. Long-term variations in signal level are caused by shadowing effects due to energy-absorbing objects (i.e. buildings and other obstacles) and short-term fading is due to multipath. The shadowing effects are mainly responsible for the hidden node problem. Consider the scenario in Figure 4.3 as an example. In this figure, primary user 1 is transmitting to primary user 2 on frequency F1. User 2 is on a hilltop and has good line-of-sight with user 1. The cognitive radio, however, cannot detect the transmission as the cognitive radio is positioned in the shadow of a large building complex. The cognitive radio therefore decides that F1 is free for use. The cognitive radio transmits on F1 and in doing so causes

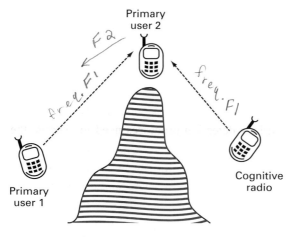

Fig. 4.3. The hidden node problem.

interference to primary user 2. Hence the hidden node problem leads the secondary user to incorrectly conclude that spectrum is unoccupied, when in fact it is not. Given the types of indoor and outdoor environments in which radios operate, hidden node problems are by no means rare. The hidden node problem is one of the most difficult problems to deal with and has to be addressed if dynamic spectrum access is to succeed. As we shall see below, a cooperative approach to spectrum sensing is seen as a potential solution to the hidden node problem as well as to some of the other limits on sensing mentioned at the outset of Section 4.4.2.

4.5 Cooperative spectrum sensing

Cooperative spectrum sensing occurs when a group of cognitive radios in some way share sensing information, in order to get a more accurate picture of current spectrum occupancy. The individual radios sense using the techniques discussed in the non-cooperative cases. But they do not rely solely on their own ability to process and detect the primary user, but instead share the burden with other secondary users in the system. We now no longer think of the cognitive radio as an individual entity and instead think of a network of cognitive nodes. The network consists of the various secondary users, looking for opportunities to use spectrum vacated by the primary users.

There are two broad approaches to cooperation. The first is a centralised approach in which some master radio (node) collects sensing information from all nodes in the network, a decision about whether the primary user is present or not is made on the basis of all of the information, and the decision is relayed back to all radios (nodes) in the network or group. The second approach is a distributed approach in which no master radio (node) exists. In this kind of system, it is possible to envisage some kind of local sharing of sensing information between, for example, one-hop neighbours. Depending on the exact mechanisms used, it may be possible for all nodes to become aware of all other nodes' sensing information or it may be that information exchange remains shared at a local level only.

4.5.1 Benefits of cooperative sensing

While there is no cooperative sensing system for cognitive radios fully operational at the time of writing, research has been carried out in this area. Though the focus of this book is not necessarily research oriented, it is impossible not to talk about cooperative sensing without mentioning some aspects of the research.

A number of different research projects have simulated the effects of multiple nodes being involved in cooperative sensing. In the main the work has focused on groups of nodes which individually deploy energy detectors and share the results of individual sensing processes. There has been a smaller amount of work that has looked at cooperation of users who employ feature detectors. The following is a list of what can be described as the benefits of cooperative sensing that more or less sum-marises the findings of the research in papers such as those by Cabric and Brodersen [3]:

- The hidden node problem can be addressed.
- An increase in the number of users cooperating leads to an increased ability to deal with worsening signal-to-noise ratios (note this means limits can be pushed further).
- Collaboration may also be used to enhance the agility of the secondary network by reducing the time required for the detection of the primary signal. This is particularly important during the ongoing secondary transmissions where fast detection of the reappearing primary users is very critical.
- Maintaining the global probabilities of missed detection and false alarm at a desired level enables users to employ less sensitive detectors, thereby reducing the hardware cost and complexity.

4.5.2 Cooperative sensing requirements

Cooperative sensing, while having benefits, comes at a price and that price is extra complexity in the network. There are extra resources required to make it possible and there will be extra work needed on the part of the individual nodes. Much of the research is directed towards getting the benefits at the least cost. The following discussion is a general discussion of the resources that can be needed.

The need for a control channel

To cooperate, radios in a cognitive network need to exchange information with each other. This is a similar problem to exchanging configuration information. In Section 3.5 the use of control channels both for rendezvous purposes as well as for subsequent exchange of information in cognitive radio was discussed. At that stage the need to exchange sensing information was not included. Whether some kind of out-of-band or in-band mechanism is deployed, or whether some kind of beaconing mechanism[5] or some technique yet to be discovered is best, remains currently unknown. As the field is so new, there is no best practice that can be followed or examples that can be cited to learn from and this remains an open problem. In Section 4.5.3 one mechanism of facilitating exchanges of spectrum sensing information in a cooperative system will be described but it is yet to be implemented commercially.

Possible need for synchronisation of sensing times

Cooperative sensing may necessitate synchronisation of activity in the cognitive network. So, for example, it may be necessary that all radios sense during one period and exchange sensing information during the next. How long the sensing period should be so that best results can be obtained is a challenge to determine. While the synchronising of networks is possible, it adds an extra level of complexity to the network. And synchronisation is very straightforward in a centralised system, and much more difficult in the decentralised case. Hence asynchronous solutions are suitable for the distributed case.

The need for mechanisms for fusing of observations

Cooperative sensing obviously involves the bringing together of different sensing results from individual radios. Consider first the centralised

---

5 A beaconing mechanism implies that information is broadcast at regular intervals on a control channel and is available to be picked up by any device choosing to listen.

method in which some kind of master node gathers and processes the sensing information. There are two methods for doing this. The first method is termed a *soft* method and in this case all radios send complete sensing data to the master node. The master node then has precise replicas of the measurements and can deduce what white spaces exist and send this information back to the nodes. This has a large overhead and would involve heavy use of the control channel. The combining of the sensing results of the various users may also be challenging as they may have different sensitivities and sensing times, and some form of weighted combining may need to be performed in order to take this into account. An alternative to this, the *hard* approach, involves the individual radios sending only the results of their individual sensing processes back to the master node. It would be possible, for example, to have a 1 or 0 to indicate whether a channel is free or not. This would cut down on the overhead. It is called a hard approach as a hard decision is passed back to the master. In the distributed scenario actual sensing information, or the hard results of the sensing process, can also be exchanged in accordance with whatever process is used (e.g. exchange between one-hop neighbours only). However, distributed sensing is not so trivial because the correlation that different users may experience may mean that not all users are independent.

Obviously this process will necessitate the use of supporting observations. For example, location details of the cognitive radio will be needed when the master node compiles a map of the combined sensing results. This may be obtained from extra equipment such as a GPS or via triangulation techniques between nodes.

The need for a suitable geographical spread of cognitive nodes

While fulfilling the requirements listed so far is not without its challenges, it is possible to find solutions. The final requirement is less easily delivered. To get the benefit of the cooperative sensing, the cognitive radios must be suitably spread throughout the geographical area that is being used. This is best explained with an example. Figure 4.4 reintroduces the

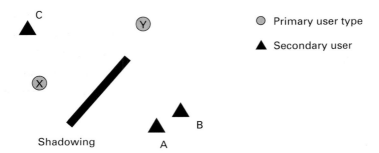

Fig. 4.4. The hidden node problem again.

hidden node problem, primary node X is shadowed from secondary user A. But if A mistakenly transmits it will interfere with primary user Y.

We have heard how cooperative sensing can help with the hidden node problem. The arrival of cognitive node B will do nothing to alleviate the problem, as it is in the same situation as node A. However, cognitive node C can contribute significantly as it can detect the hidden node. Hence the cooperation between A, B and C can lead to a clearer picture. The main point to take from this is that, while cooperation can aid significantly in the sensing process, it is only useful provided the cognitive radios are suitably distributed. While this might seem a difficult challenge to meet, it is clear that, as network densities increase, meeting this requirement becomes more likely.

4.5.3 A cooperative sensing example

We now turn our attention to an example of a proposed cooperative sensing system in the unlicensed use of the TV white spaces. In Chapter 1 the move to digital TV was discussed and the proposal by the FCC to make the TV white spaces available to unlicensed users was introduced. As a result of this kind of initiative the IEEE 802.22 working group was set up to develop a standard for wireless regional area networks that would make use, on a non-interfering basis, of the TV white spaces. We will briefly look at the overall standard before focusing specifically on the sensing aspects of the standard.

The IEEE 802.22 standard

As stated already, 802.22 is a standard that will describe a physical and media access layer of a wireless regional area network that is intended to make use, on a non-interfering basis, of unused TV broadcast channels of 6 MHz, 7 MHz or 8 MHz in bandwidth. TV channels in the USA are 6 MHz in bandwidth. The 7 MHz and 8 MHz are to facilitate other jurisdictions. The envisaged usage of these networks is for the provision of unlicensed broadband services, particularly in rural and remote areas. The TV band frequencies make them very attractive for services that require long distances. The 802.22 specification is still being debated and discussed; an opening report was produced in July 2008. While there are many issues yet to be decided, the broad outline of the specification is well known. Very good descriptions of the 802.22 standard, in so far as it has been developed, can be found in the publications by Cordeiro *et al.* [4].

802.22 specifies that the network should operate in a point to multipoint basis. A basestation which is a professionally installed entity will manage its own cell in which a number of consumer premise equipments (CPEs) operate. Hence there will be infrastructure in place to help unlicensed users avail of the TV white spaces. Currently when we use the existing unlicensed bands there are situations in which we put infrastructure in place as well. An example is the case of WiFi systems. In this case access points need to be installed. These access points are like small basestations themselves and manage the users within their range. However, access points are small, relatively low powered and relatively cheap, and the access points are guaranteed to have access to the frequencies associated with the ISM band available to them. The TV white spaces open up a very new way of thinking in that more costly and larger infrastructure will be put in place, in order to opportunistically avail of whatever white space may (or may not) exist.[6] In the case of 802.22 it is envisaged that the basestation will control access to whatever spectrum is available for use. To do this all entities in these networks must have cognitive capabilities, i.e. it is a cognitive radio network.

---

6 However, it is guaranteed that white spaces will exist, so it is not the case that capital
  expenditure would take place and no white spaces be in existence.

Sensing in IEEE 802.22

There are many interesting aspects to the IEEE 802.22 standard and it has huge relevance throughout the book. However, for the purposes of the current chapter, we focus only on the sensing aspects of the standard. The whole aim of the specification is to find a mechanism that allows the unlicensed users to make use of the TV white spaces without causing interference to the incumbents (TV stations and wireless microphones). Hence the ability to sense incumbents is core to its success.

All CPEs in an 802.22 system must have the ability to perform spectrum sensing. The standard will not specify what kind of sensing technique is to be used. Instead it specifies that all CPEs must be able to detect signals: digital TV signals that are as low as -116 dBm, analogue TV signals that are as low as -94 dBm and wireless microphone signals that are as low as -107 dBm. The designers of CPEs for 802.22 operation can make whatever software and hardware choices they wish, and whatever tradeoffs they see fit, to deliver on this requirement.

While the standard will not proscribe an approach to sensing, a number of sensing techniques have been explored in the context of 802.22 (all of which assumed that there will be a separate sensing receiver). The various suggestions can be classified under the three headings addressed already in this chapter, namely energy detectors, matched filter approaches and feature detectors. Both analogue and digital TV channels have relatively unique spectrum signatures that are easy for cognitive radios to identify, and a number of the suggested feature detector approaches focus on various unique aspects of the TV signals. Suitable sensing approaches will be listed in an appendix of the standard.

Sensing has been subclassified further within the standard and two levels of sensing have been defined, namely fast sensing and fine sensing.

- Fast sensing: The requirement here is to sense the channels of interest at under 1 ms per channel. Typically energy detection is envisaged for this. It gives a quick and blunt view of what is happening in a channel.
- Fine sensing: This involves more detailed sampling and can take the order of milliseconds per single frequency channel. A feature detector would typically be used at this stage.

All of the techniques for sensing that are explored in the context of 802.22 will be classified as either fast or fine.

In terms of the hardware associated with sensing, apart from an expectation that a separate receiver would be used for the sensing process there is also an expectation of there being a separate antenna. The IEEE 802.22 standard is based on using two antennas, a directional antenna for communication and an omnidirectional antenna for sensing. The directional antenna for communication between basestation and CPE can help avoid interfering with incumbents. The omnidirectional antenna is essential to sense incumbents from all directions.

Cooperative sensing in IEEE 802.22

We have chosen to discuss 802.22 to illustrate cooperative sensing. Thus far we have simply described the individual sensing capabilities of the CPEs in the 802.22 network. The cooperation comes into play via the basestation as the basestation coordinates all the CPEs, controls the sensing process and gathers the results of the sensing process together. This allows the basestation to create a list of occupied and unoccupied channels, or in other words a spectrum map which can be used to determine what white spaces are available and what CPEs should use which channels. If, as a result of the sensing process, a basestation realises that it is operating on the same channel as an incumbent user it must broadcast a message to all users in the service area and switch to an unoccupied channel within a 2 second time period. The gathering of the sensing information to create the global spectrum occupancy map is not prescribed in the standard. This means that solutions which involve hard fusion of information or soft fusion of information, as discussed in Section 4.5.2, can be employed.

A few more details of the sensing process can shed more light on the cooperation. The basestation directs the CPEs *when* to sense as well as giving instructions about *what* (channels) to sense. To understand how this happens it is necessary to introduce the notion of in-band and out-of-band sensing.

- In-band sensing: This refers to the sensing that takes place in the frequency channels on which the basestation is communicating with the CPEs.
- Out-of-band sensing: This refers to sensing that takes place on all other channels.

We have already discussed that sensing not only involves spotting the initial opportunity to use some available white space, but also involves continuous monitoring of the spectrum subsequently to detect the return of the primary user. The in-band sensing can be considered to be looking for the return of primary users while the out-of-band sensing can be seen to be associated with looking for alternative free channels to use, should the current channels become unusable or insufficient.

In terms of the in-band sensing the basestation periodically quietens all CPEs and then incumbent sensing can take place during these quiet periods. Both fast and fine sensing can be used. In terms of out-of-band sensing, again both fast and fine sensing can occur on all the remaining channels. The basestation dictates whether fast or fine sensing is used. Fast sensing could be carried out on all channels and then a more detailed level of sensing (fine sensing) applied to a subsection of channels where further observation was deemed necessary.

4.6 Getting information from an external source

In the introduction to this chapter it was emphasised that cognitive radios could, rather than perform sensing themselves, turn to other sources for information, or supplement the sensing they perform with additional information provided by some service. In broad terms there are two possible ways in which such a service can manifest itself. It could manifest itself as some widely accessible database or as a physical infrastructure network whose sole function is sensing of spectrum occupancy.

4.6.1 The accessible database

A regulator – a spectrum manager, the Government or some such entity – may make an electronic database available that contains information about users of spectrum. Therefore rather than perform sensing itself a cognitive radio would consult the database in order to determine what spectrum is free. This kind of approach is recommended by the FCC in their 2008 report, which outlines rules for unlicensed use of the TV white spaces. There are those who feel it may produce a more reliable output than allowing radios to sense themselves. The database could contain locations of TV transmitters, details of the frequencies on which they transmit, duration of transmissions, characteristics of the transmitters such as transmit power, etc.

A database has the advantage that it can contain much more than transmitter information. More detailed information about the landscape and environment can be included (for propagation modelling). In the next chapter we discuss how regulatory polices can be stored in a database. So it may be the case that, even if a cognitive radio does sensing itself, it will consult a database for other information such as details about regulations in the jurisdiction it currently finds itself.

There are negatives to a database, especially in the spectrum sensing context. A database may be very static and not offer the potential for cognitive radios to exploit variations in the transmitted signals over short time spans. And, even more importantly, the sensing database is discussed here solely in the dynamic spectrum access primary user/secondary user context. As we have already mentioned a number of times, cognitive radio has far wider applications. Sensing may be needed to coexist with neighbours in many other applications described in Chapter 1. These other applications cannot use a database and very much rely on *in situ* sensing.

4.6.2 The service network

It is also an option to build a large-scale network that performs sensing and operates as a service from which information is bought. On one level this is like a private database. But the main difference here is that live information is fed to the database from the sensing network. While this might sound somewhat far-fetched, spectrum monitoring, as will be shown in Chapter 8, is a service needed by regulators and the type of service advocated here is akin to that.

The service network leads to another important topic. It should be possible to determine the cumulative level of interference that exists at any one time in a given location. This would open the way for a much more sensitive analysis of 'whether there is room' for another transmission. This brings up the issue of *interference temperature*.

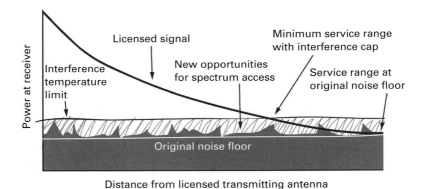

Fig. 4.5. The interference temperature explained.

The US Federal Communications Commission, in 2002, chartered an internal staff study, the Spectrum Policy Task Force, to investigate the future needs of RF spectrum and the limitations of current spectrum policies, as well as develop recommendations for enhancing current policies. The suggestion of a common interference metric was one of the outcomes of the task force. The proposed metric was termed the interference temperature. The Notice of Proposed Rule Making, ET Docket No. 03-237, details the ideas involved. Figure 4.5, reproduced from the Notice of Proposed Rule Making, is often used to portray the interference temperature idea. The figure shows the power of a signal at a receiver, which drops off with distance from the transmitter. There is a distance beyond which a receiver will not be able to make sense of the signal. Another way of saying this is that the signal power becomes so low that it is indistinguishable from the noise floor. In Figure 4.5 the point at which this happens is marked on the very right of the image. Any interference that occurs that increases the noise floor will mean that the receiver becomes unable to function correctly. Such interference could be the result of a secondary user. However, suppose we allow for a certain amount of interference, albeit resulting in a reduction in range at which the transmitted signal can be properly picked up, then we open the possibility of other transmissions being allowed to take place. In other words we stipulate that, provided the noise floor does not go above a certain level, other transmissions are possible. The higher noise floor level is indicated in the diagram and the opportunities for spectrum access are highlighted. In other words the interference metric can be used to determine underlay opportunities.

The interference temperature is a measure of the RF power available at a receiving antenna as a result of all the signals impinging on it. The reason the word 'temperature' is used is because the temperature equivalent of the RF power available at a receiving antenna per unit bandwidth, measured in Kelvins, is used. The idea of creating such a metric is to be able to make statements about *how hot* a receiver can get. It does not just apply to determining opportunity for underlay. Secondary users can transmit if the temperature at the primary receivers, as a consequence of the secondary users' transmission, does not exceed a certain threshold, i.e. if it does not get too hot at the primary receivers.

There has been some progress on the development of models for and approaches to the measurement of interference temperature. However, despite this, the FCC in May 2007 came to the following conclusions:

Commenting parties generally argued that the interference temperature approach is not a workable concept and would result in increased interference in the frequency bands where it would be used. 1. While there was some support in the record for adopting an interference temperature approach, no parties provided information on specific technical rules that we could adopt to implement it. 2. Further, with the passage of time, the Notice and the record in this proceeding have become outdated. We are therefore terminating this proceeding without prejudice to its substantive merits. Accordingly, IT IS ORDERED that, pursuant to Sections 4(i) and 4(j) of the Communications Act, 47 U.S.C. 154(i) and 154(j), ET Docket No. 03-237 IS TERMINATED, effective upon issuance of this Order.

It is at this point we finally return to the service network. Interference temperature is a very attractive metric. One way to make it workable might be through a network of sensors, such as the service network described here. An understanding of the spatio-temporal nature of interference temperature would be needed. This could help determine the density of the network that is needed to get reliable or good enough results as well as be used to interpolate between the sensed locations to get interference temperature at a specific point.

4.7 Back to the wider observations

The chapter began with a very broad description of the inputs that a cognitive radio uses. However, after the initial broad discussion, there followed a focus on the much narrower topic of spectrum sensing. While sensing is of major importance, there are many more observations that need to be made about the radio environment. More information is particularly useful when guiding the actions the radio takes. We therefore briefly broaden the conversation again.

Once communication between two nodes is established there is a whole range of observations that can be made. These include received signal strength, bit error rate, packet error rate, packet delay time, signal-to-noise ratio at the receiver and channel impulse response, to name a few. Most of these metrics have already been explained in Chapter 3. *Packet error rate* has not come up before and is simply the ratio of incorrect packets received to the total number of packets sent. The *packet delay time* has also not been used and is self-explanatory. *Channel impulse response* is another new term and this needs more explanation. Channel impulse response is a measurable response by a radio communication channel when an electromagnetic impulse is transmitted over the air. The transmitted signal, as was discussed in Chapter 3, can experience multipath and other effects. To the signal this looks as if the radio propagation channel is acting as a time and spatial varying filter. If these filter characteristics can be determined, then it is possible to understand something about the propagation environment as well as to use the information to take corrective action at the receiver. The purpose of multiple impulse measurements is to quantify the varying filter characteristics of the communication channel in both the time and spatial domain. Channel impulse response can be measured using many different techniques. One example involves the use of known training sequences. A receiver will know what to expect and can compare expected values with the actual received values and use this information to determine the channel impulse response.

There are other techniques that do not need the transmitter to use training sequences or pilots; these are referred to as blind channel estimation techniques.

The radio can make more network-oriented observations. This has been alluded to already. These can include observations about the number of nodes in the network, the network topology, the speed of the nodes, the traffic flow patterns, etc. The radio can also make observations about its own performance such as the percentage of central processing unit (CPU) usage or power expenditure or battery levels.

The list is endless. Rather than continue expanding the list, it is better to make a few important but general points.

1. Many of the measurements referred to here occur in different existing communication systems. This is very welcome, because it means that well-used techniques already exist to make the many observations that are of relevance to a cognitive radio.
2. A cognitive radio should attempt to make as many of these measurements/observations as possible. The richer the picture it can paint, the better placed it is to decide on what actions should be taken.
3. Multiple techniques may be needed for some measures as the technique may depend on the waveform in use. For example multi-carrier communication systems will lend themselves to certain kinds of approaches to channel estimation.
4. Measurements or observations need to be made widely available in the radio (and among radios). This means making the measurements available to all layers and components of the cognitive radio system. It also means that the collaboration discussed in the context of spectrum sensing can also apply to sharing other measurements and fusing the shared observations among cognitive radios.

4.8 Conclusions

This chapter on observing the outside world has mainly focused on how spectrum sensing is carried out. It is clear that making observations relating to spectrum occupancy is a core function of a cognitive radio and, as stated at the outset of the chapter, the observe stage of the cognitive radio is often synonymous with the act of spectrum sensing. Much progress has been made in spectrum sensing but issues remain that need to be addressed, particularly in the context of cooperative sensing and the development of interference metrics. The option for using external sources of information is a very valid option. And databases can hold much more than spectrum occupancy information.

There are many other observations which a radio can make. Many well-understood and robust measurement techniques already exist that can be harnessed and included in a cognitive radio. These observations are needed to guide the wide choice of actions detailed in Chapter 3. The word 'awareness' is often used when describing a cognitive radio. Awareness is very much dependent on a rich flow of observations.

References

1. S. Haykin, Cognitive dynamic systems, *Proceedings of the IEEE*, **94**:11 (2006), 1910–11.
2. F. Seelig, A description of the August 2006 XG demonstrations at Fort A.P. Hill, in *2nd IEEE International Symposium on New Frontiers in Dynamic Spectrum Access Networks, 2007*. Dublin, 17–20 April 2007, pp. 1–12.
3. D. Cabric, and R.W. Brodersen, Physical layer design issues unique to cognitive radio systems, in *16th IEEE International Symposium on Personal, Indoor and Mobile Radio Communications (PIMRC), 2005*. Berlin, 11–14 September 2005, **2**, pp. 759–63.
4. C. Cordeiro, K. Challapali, D. Birru and S. Shankar, IEEE 802.22: An introduction to the first wireless standard based on cognitive radios, *Journal of Communications*, **1**:1 (2006).

5   Making decisions

5.1 Introduction

We now reach the 'decide' part of the 'observe, decide and act' cycle. In very simple terms the decision-making process is about selecting the actions the cognitive radio should take. Using the vocabulary introduced in Chapter 2, it is about choosing which 'knobs' to change and choosing what the new settings of those 'knobs' should be. Decision-making goes very much to the heart of a cognitive radio.

5.2 The decision-making process: part 1

In Table 3.2 a variety of cognitive radio applications and the main *high-level actions* associated with them were presented. On examining the table we noted that many of the actions, whether commercial, public safety or military based, centre on two activities:

1. The cognitive radio shapes its transmission profile and configures any other relevant radio parameters to make best use of the resources it has been given or identified for itself, while at the same time not impinging on the resources of others.
2. If and when those resources change, it reshapes its transmission profile and reconfigures any other relevant operating parameters, and in doing so it redirects resources around the network.

A re-examination of Table 3.2 will confirm that these actions are standard throughout a whole variety of applications. It therefore comes as no surprise that two kinds of decisions that regularly need to be made are decisions that map to these two activities, namely decisions about how resources are distributed and decisions about how those resources are exactly used.

The decision about 'who gets what resources' is a very fundamental decision in cognitive radio. In many application areas, resources such as spectrum may be assigned by a third party such as a regulator. However, there is a range of application areas in which more dynamic decisions need to be made about resources. For example, in dynamic spectrum access systems, once unoccupied spectrum has been identified, that spectrum may need to be divided among access points who will serve customers with the spectrum. Likewise in a public safety scenario, as needs change, decisions may have to be made about where spectrum resources should be targeted. If resources can be equated to a cake, then the cake must be divided up in some manner. There is usually some higher-level motivation for dividing the cake, such as 'everyone gets their fair share' or 'hungriest get more'. Hence questions like 'what is the optimal way to share out resources that is workable from an interference perspective and ensures fairness?' or 'what is the optimal way to ensure those who need resources most get them?' are asked. These are *optimisation problems*.

Once resources are made available to individual nodes, the radio can then accordingly shape its transmission. This high-level action can be broken down in a series of (sub)actions. As pointed out in Chapter 3, the signal can be shaped to use resources by manipulating its frequency content, its spatial footprint and its temporal profile, and the available resources can be exploited by making sure the signal is robust for its journey and makes optimal use of available capacity. Each of these actions themselves can be further broken down into a set of smaller actions, many of which were discussed in Chapter 3. So the decision-making process is about selecting exactly which actions or sub-actions should be taken by the radio. As we saw in Chapter 3, many actions can have opposing or contradictory effects and hence some effort is needed to seek out the exact combination that produces the desired output.[1] Even though the choice of a number of actions may be restricted or constrained by some physical bounds, there may be many combinations of actions which can deliver the kind of transmission profile that is of interest. Looked at

---

1 This description is not meant to suggest that choice of every action is the radio's sole responsibility. Many actions can only be taken in concert with neighbours. Recall the unilateral and multilateral decision-making processes introduced in Chapter 2.

in this manner, the essential element of the decision-making process is yet again an *optimisation problem*. It is about finding a combination of settings (for the knobs) that gives the desired transmission profile (or delivers whatever other high-level action is of interest) or, failing that, something as near as possible to the desired profile.

Hence the key point of our first exploration of decision-making is that optimisation is at the core of this process. We begin by looking at this core activity and subsequently build up a description of the other processes which play a role.

5.2.1 A walk on the optimisation landscape

An optimisation process can be defined as the process involved in selecting the 'best' choice from the list of available choices in order to reach some kind of goal or at least get as near as possible to the goal. Therefore we need to know (i) the goal, (ii) what choices are available to reach the goal and (iii) what 'best' means.

The goal
In the decision-making process, the word *objective* is used to describe the goal that is of interest. The phrase *objective function* typically refers to a mathematical expression for the goal. Often the objective function is expressed in terms of maximising or minimising some metric. So, for example, when transmitting a signal the aim might be to *maximise* throughput for a given bit error rate. Or in the case of deciding who gets what resources, the aim may be to *minimise* inefficiencies or *minimise* occurrences of unused spectrum. It is also possible to have a goal with multiple objectives. So, for example, a radio might want to *maximise* throughput AND *minimise* the amount of power used. The multiple objective case is very common in cognitive radio applications. The radio or network of radios wants to achieve the goal under a given set of conditions; the conditions are understood via observations that are made.[2]

---

2 Recall that the term 'observation' as used in this book applies to all the inputs gathered by the radio that lead to an understanding of its environment and the circumstances in which it finds itself.

The choices

The radio must choose which actions to take to reach the goal. As mentioned already, there can be a wide range of actions which suit. We focus on a very simple example for the purposes of this exercise and look at maximising throughput by selecting an optimal modulation scheme. The list of choices can include, for example, use BPSK, use QPSK, use 8PSK, use 16QAM or use 64QAM.[3] Some of these actions may be not be possible. For example suppose the radio cannot do 64QAM modulation. This means that the list can be reduced. A similar analysis can be brought to bear on decisions about resources. Suppose now we have three access points A, B and C and three frequencies to distribute F1, F2 and F3. If an assignment is labelled by radio name and frequency then we have the following possible distributions (A-F1, B-F2, C-F3) or (A-F3, B-F1, C-F2) or (A-F2, B-F3, C-F1). If, for example, radio B can't deal with frequency F3[4] then only the first two are valid. In optimisation a list of choices (or solutions) such as the examples given here, is called the *search space*. The term *solution* or *candidate solution* is usually used to refer to each of the choices on the list (or in the search space).

Determining 'best'

Determining 'best' means evaluating all the choices or candidate solutions (action or a combination of actions) to see which one ranks highest. In the throughput example 'best' means the modulation scheme which gives the maximum throughput in the prevailing conditions. To determine this we need to be able to express the relationship between the throughput and the action (setting of modulation scheme) for the prevailing conditions. In this case relevant information about the prevailing conditions is gleaned from observing the signal-to-noise ratio at the receiver. This observation can be fed into the expression for throughput and modulation scheme, and hence the modulation scheme chosen will be the one that maximises the throughput for a given bit error rate. The term *measure of fitness* is used to express how well the

---

3 These are the names of various modulation schemes. The larger the number the more compact the data is as it is transmitted and hence the higher the throughput.
4 Perhaps it does not have the correct RF frontend for F3.

particular solution does. When multiple entities are involved, for example in decisions about the distribution of resources, it will be an overall measure of fitness that is taken into account rather that the measure of fitness to an individual radio/user. A simple measure in the example of frequency assignment to access points is the number of radios satisfied.

Searching the landscape

In the discussion thus far very simple examples have been used. In both the maximising throughput example and assignment of frequencies example, there is a finite and manageable set of actions to consider. The optimisation problem is in fact a *deterministic*[5] one as there is a clear relationship between the goal and the choices that can be made to reach the goal. In reality, however, the search can be very large and possibly infinite and the relationship between the goal and the actions may not be at all clear.

Suppose now we take a more complex view of how to maximise the throughput and discuss the issues that can arise. The following longer list of actions can be created:

- use a higher-order modulation scheme
- use antenna diversity techniques
- increase the transmitter power
- use better compression techniques
- use lighter error coding (i.e. have fewer redundant bits).

As is evident from the list, we now have the possibility of using all sorts of characteristics of the radio to achieve our goal. We are actually carrying out a *cross-layer optimisation* as the optimisation now happens, as the name implies, 'across layers' (PHY and MAC) of the radio. It is possible to imagine the search space getting very large if this list were properly expanded to include all types of modulation schemes, coding schemes, compression techniques, power ranges and every combination thereof.

---

5 Deterministic means that everything that happens is determined by a necessary chain of causation.

Again it may be possible to rule out certain actions. For example, it may not be possible to increase power without causing harm to a neighbour or, put more generically, without violating some constraint. This in turn may rule out certain modulation techniques, which will be dependent on certain levels of quality of the signal, which may also affect coding and compression choices. However, despite all this, there may still be a large search space. If a search space is large or infinite it is not possible to search it all and only a partial search is possible.

To understand the search process we need some further discussion. The aim of the optimisation process is to find a maximum or minimum of some objective function. The term *global maximum* or *global minimum* is the overall maximum or minimum in the search space. There may be other points which seem like maxima or minima but in fact are not. They are merely *local maxima* or *local minima*. The optimisation process can be equated with the search of a physical landscape for some feature. Suppose I want to find the lowest valley (equivalent to minimising our throughput) on the landscape. I can look around me and head off in a 'downwards' direction and come to a deep valley. However, I may not realise much deeper valleys exist over the other side of the mountains. I am therefore only standing on a local minimum and not a global one! In radio terms I may have found a combination of actions that minimise my throughput but there is a better combination out there somewhere. The point of the many search techniques that have been designed in the past decades is basically to come up with strategies to make it more likely to find the global maximum or minimum.

Certain landscapes may be more difficult to search than others. The kind of landscapes that can exist are depicted in Figure 5.1. This figure is taken from an excellent ebook on optimisation by Thomas Weise [1]. Some of the landscapes have many local minima, making them difficult to deal with. Other have very steep local minima and, continuing the analogy of valleys, it may be very easy 'to get stuck in the wrong one'. The concept of *gradient* is an important one in this discussion. Let us go back to the search for the valley. If I take a step and end up in a position that is higher than my previous position, I am not going in the

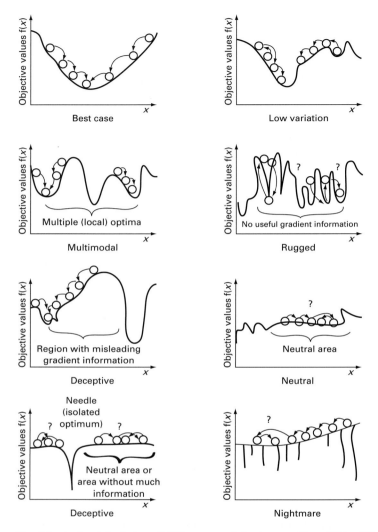

Fig. 5.1. A visual depiction of different search landscapes. From Weise [1].

correct direction. I am heading uphill. So the gradient acts as an indicator. However, it is possible that the gradient may also increase because you stepped on an isolated mound. The gradients are shown in Figure 5.1. As can be seen sometimes the gradient gives a great indication of whether you are heading in the correct direction while at other times it can give misleading or simply no information (e.g. when on the flat).

It is possible that the global minima (or maxima) may never be found. Because of this the many optimisation techniques that exist need termination criteria. Termination criteria can include:

1. Time limits: The process stops after time t. A general rule of thumb is that you can gain improvements in accuracy of optimisation only by investing more time.
2. Iteration limits: The process stops after N iterations.
3. Good enough result obtained: The process stops once some predefined notion of good has been achieved.
4. Good enough improvement: The process stops after the percentage improvement is not getting big enough with each step (i.e. the gradient is no longer decreasing/increasing significantly with each step).

In a cognitive radio there may be different reasons for applying these termination criteria. For example, in the case of dynamic spectrum access there is a time factor at play. Decisions must be made about how to shape transmissions in a timely fashion. Otherwise the transmission opportunity may pass. It may be possible to do some optimisation offline rather than online, depending on the application. A cognitive radio may have different scales and meanings for the concept of 'good enough'. There may be a high bar for good enough when it comes to interference and coexistence related issues and a much lower bar for 'good enough' when it comes to throughput, for example. The stopping criterion must therefore be matched to the task at hand.

In summary, optimisation is about maximising or minimising some objective function. In the context of cognitive radio this involves looking through a search space to find the action or combination of actions that are needed to maximise or minimise the desired objective function. In many of the types of optimisation problems that feature in cognitive radio, the search space will be large and probabilistic methods will be needed. The challenges call for techniques that can effectively search the landscape of interest.

Metaheuristics

Heuristics and *Metaheuristic* techniques provide answers to the challenges of large search spaces. The term heuristics applies to approximate optimisation techniques (they do not search all the landscape) that allow good solutions to be found more quickly.[6] Metaheuristics are more general optimisation algorithms or black box procedures for solving these computationally hard problems. Metaheuristics can be applied to wide ranges of problems rather than designed for very specific optimisation problems. This makes metaheuristics very suitable for cognitive radios. What make metaheuristics even better is that metaheuristic approaches involve using tricks so, to put it simply, you don't 'get stuck in a rut', i.e. stuck around local minima or maxima. To appreciate this concept, think of the landscape surfaces in Figure 5.1 consisting of rubber and think of pulling and stretching the rubber landscape to get a ball to move out of a local minimum. This is an analogy of the trick of a metaheuristic.

There are very many different kinds of metaheuristics. The various techniques have different ways of searching through the search space and have different tricks for not getting stuck. Metaheuristics, let alone optimisation and many of the other topics in this chapter, are whole fields in themselves and each would warrant a separate book. Hence this chapter cannot really do any of them justice. It is a recurring theme of cognitive radio that so many different fields are relevant. This makes cognitive radio simultaneously exciting and challenging. The purpose of this current section and others that follow is merely to flag the existence of techniques and approaches, highlighting a few select points.

Some examples of metaheuristics include *hill climbing* searches, *greedy algorithm* searches, *tabu* searches, *simulated annealing* searches,

---

6 From the Oxford English Dictionary: 'A heuristic process or method for attempting the solution of a problem; a rule or item of information used in such a process. A process that may solve a given problem, but offers no guarantees of doing so, is called a heuristic for that problem . . . ' For the moment, we shall consider that a heuristic method (or a heuristic, to use the noun form) is a procedure that may lead us by a short cut to the goal we seek or it may lead us down a blind alley . . .

genetic algorithm and other evolutionary programming methods. The landscape analogy used earlier helps in describing the techniques (still assume we are looking for a local minimum). The greedy search just advances downhill, quickly getting to a point that minimises the objective function and delivering a solution but perhaps not the best. The tabu search brings more sophistication and has a mechanism for declaring solution candidates which have already been visited as tabu. This means uphill moves can occur if all downhill are tabu. The simplest realisation of this approach is to use a tabu list which stores all solution candidates that have already been tested. Simulated annealing also allows for downhill and uphill moves. Simulated annealing is a technique that is inspired from annealing in metallurgy, a technique involving heating and controlled cooling of a material to increase the size of its crystals and reduce their defects. The heat causes the atoms to become unstuck from their initial positions and wander randomly through states of higher energy while the slow cooling gives them more chances of finding configurations with lower internal energy than the initial one. If we allow the system to become very hot we can move around randomly in our search and move uphill as it were. If we decrease the temperature we can move downhill and focus in on the goal. Genetic algorithms draw inspiration from evolution and use totally different techniques again. Genetic algorithms take a population of chromosomes using a genetic operation called *selection* and mix the genes of its members through a genetic operation called *crossover* to produce offspring. The chromosomes model the problem at hand in some manner. The offspring solutions can be further altered using a genetic operation called *mutation* in a random fashion. This really opens up the search. Each member of the entire population is then evaluated for 'best' fit as discussed already. Good chromosomes survive and are reproduced and the rest are discarded. Because only the best-performing combinations are permitted to survive, and those combinations reproduce further, it is possible to yield progressively better results as the process evolves. Genetic algorithms tend to thrive in an environment in which there is a very large set of candidate solutions and in which the search space is uneven and has many hills and valleys. They will be greatly outclassed by more situation-specific algorithms in the simpler search spaces.

Multilateral decisions and distributed searching

Many of the decisions in a cognitive radio are multilateral in nature. This is especially the case for decisions about shared resources (i.e. like the time-share apartment). We can discuss this further with an example. Consider Figure 5.2, which shows a number of cognitive nodes. Suppose the task at hand is similar to the resource optimisation task given as an example earlier in the chapter. In that example we had three access points A, B and C and three frequencies to assign, namely F1, F2 and F3. Now we have N radios and M frequencies to assign. In this case M is less than N so frequencies have to be reused. Figure 5.2 also features some annotations which list the restrictions that might apply to any assignment. We need to search through the various combinations of assignments while bearing these restrictions in mind.

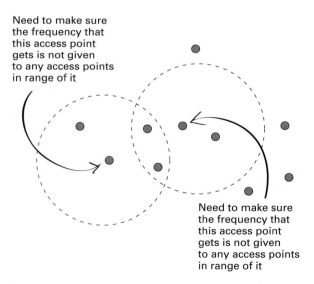

Need to make sure the frequency that this access point gets is not given to any access points in range of it

Need to make sure the frequency that this access point gets is not given to any access points in range of it

Fig. 5.2. A set of N access points which must each be assigned a frequency from a group of $M < N$ frequencies. Two nodes are annotated as examples of the restrictions on them.

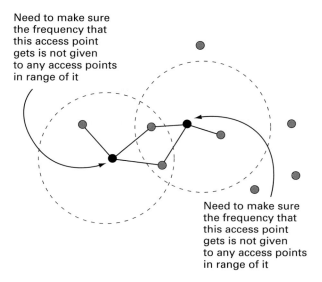

Need to make sure
the frequency that
this access point
gets is not given
to any access points
in range of it

Need to make sure
the frequency that
this access point
gets is not given
to any access points
in range of it

Fig. 5.3. A graph representing constraints is beginning to be formed around the two sample nodes.

In order to take restrictions, such as those illustrated in Figure 5.2, into account, some kind of *network-like model* that captures the interdependencies between the nodes is needed. The purpose of this is to create a formal representation of the interdependencies to allow the problem to be tackled mathematically. One way to do this is to consider the network of nodes as a graph. A graph refers to a collection of vertices and a collection of edges that connect pairs of vertices. In Figure 5.3 a graph is being created. Two nodes have an edge between them if they interfere with each other and hence cannot be assigned the same frequencies.[7] The graph shown here is only partially complete, for illustration purposes. The

7 Interference occurs in the diagram here if two nodes are within interference range of each other. In the diagram a circle is drawn around the two access points that are used as examples and the nodes within the circle experience interference. In reality the area of interference will not be a neat circle – instead it will be more like a paint splat! In reality also some kind of observation process will be needed to determine who is in danger of interfering with whom. This is another process that may call for the use of some kind of control channel which would be used to make measurements to determine whether a node was within the interference range of another node or not.

remaining interdependencies need to be filled in. The optimisation now becomes what is known as a *graph colouring* problem. The vertices of the graph must be coloured (i.e. assigned a frequency) such that no two adjacent vertices share the same colour (frequency). The search algorithm used will attempt to find combinations of colours that suit. A good measure of success is that all access points get assigned a frequency or that the least number of frequencies possible is used in the colouring task. The graph colouring approach is only one such way of taking interdependencies into account. There are Bayesian network techniques and other constraint programming techniques that set up the problem and allow for searches of the optimal assignment combinations.

Once the problem has been set up, the optimisation process or search can begin. For centralised decision-making processes the metaheuristic techniques described so far suffice. For decentralised cases, distributed or what are called parallel versions of the techniques are needed. The great challenge in distributed optimisation problems is how to reach a global network-wide optimum when working at a local individual node level only, i.e. without the god-like view of the centralised approach. There are distributed versions of some of the metaheuristics such as parallel genetic algorithms or parallel tabu searches. *Island Genetic Algorithm* is an example of one such programme. In this case the genetic algorithm runs on each node and periodically the best solutions migrate between nodes. A nice summary of the different techniques can be found in Chapter 9 of Friend *et al.* [2]. It should be noted, however, that optimisation problems that are distributed are much more challenging than ones that are either centralised or focused on unilateral decisions only. And there are many open research problems to be tackled.

There is an important topic that should also be mentioned in the context of distributed optimisation and that is *game theory*. Stated simply, game theory is a collection of models and tools for analysing *interactive decision problems*. An interactive decision process is a process whose outcome is a function of the inputs from several different decision-makers who may have potentially conflicting objectives with regard to the outcome of the process. Multilateral distributed decisions, such as those we are dealing with in this section, are interactive decision problems. The decision made by one person affects the others.

In the discussion thus far we have spoken about applying distributed optimisation algorithms, albeit without explicitly mentioning it, in a *cooperative* fashion. We assume that the overall goal of the network (i.e. what is best in general) will be what each node in the network works towards. So it will take fewer resources if it means more people will be happy. This is not necessarily what happens in real life and there are many cases of *uncooperative* behaviour. Recall that in the taxonomy of decision types presented in Figure 2.4 uncooperative distributed decisions featured. The time-share decision in which nobody communicated about holiday plans was used as an example. If you want to avoid sharing the apartment with some unbearable friends there may be incentives to cooperate! In radio terms, decisions about power settings, spectrum usage or routing choices are similar. Without cooperation everyone can end up using excessive power to try to drown each other out, everyone can converge on the same empty spectrum or everyone can use the same routes and cause congestion. Game theory provides a framework for studying both cooperative and uncooperative interactive decisions.

Every game (interactive decision) includes a set of players, actions for each of the players, some method for determining outcomes according to the actions chosen by the players, preferences for each of the players defined over all the possible outcomes and rules specific to the model. In the case of cognitive radios, the radios are the players in the game. There are many different kinds of game, for example, games where players have one go only, games where players have multiple goes, etc. The main point of game theory is to find equilibria points in the games. These are sets of strategies in which individuals are unlikely to change their behaviour. Many equilibrium concepts have been developed, the most famous of which is the Nash equilibrium.[8] The Nash equilibrium states that, if each player has chosen a strategy and no player can benefit by changing his

8 John Forbes Nash is an American mathematician who works in game theory and won a Nobel Prize for his efforts in the field in 1994.

or her strategy while the other players keep theirs unchanged, the current set of strategy choices and the corresponding payoffs constitute a Nash equilibrium. A strategy here would correspond to a power level choice, a frequency band choice, a routing choice, etc. There are other equilibria which can be used as well. The point here is that game theory allows the study of different strategies that the radios can embrace to gain a deeper understanding of how the network will react and behave. While game theory is essentially a means of gaining insight and understanding, it is possible also that the understanding will lead to the design of incentive mechanisms that can be used to encourage the kind of local behaviour that leads to what is globally 'best' for the network of cognitive radios.

5.2.2 Summary part 1

The main message here is that optimisation is at the core of the decision-making process. We saw that some problems lend themselves well to deterministic approaches but many more need probabilistic techniques because of the very large and even infinite complex search spaces that are involved. Metaheuristic approaches provide a very manageable way of dealing with the many optimisation problems that crop up in cognitive radios. There are a wide variety of techniques to choose from and it is likely that an advanced cognitive radio would have a suite of optimisation tools to hand. Dealing with centralised problems is less challenging than distributed problems so a cognitive basestation that optimises resources for the nodes it serves has an easier task than an ad hoc network that is optimising itself in a distributed manner. Having said that, the field of optimisation is well advanced and it is a case of selecting appropriate techniques that are suited for use at time scales that are commensurate with the decisions that need to be made. Providing some of the information for the optimisation problems is perhaps a bigger challenge than finding a suitable optimisation technique for the decision-making process at hand. This is particulary the case when dealing with distributed resource allocation decisions in which control channels or some such entity may be called for yet again to exchange relevant information.

5.3 The decision-making process: part 2

Decision-making is not all about optimisation. In fact there is much work to be done above and beyond the optimisation process. To illuminate this point we work again with the decisions involved in the high-level action of shaping a transmitted signal. Figure 5.4 paints a picture of what can happen and the following points elaborate further.

1. Firstly the decision-making process has to be *triggered* into action. The process can start as a part of start-up of the radio, when a radio gets allocated new resources, when neighbouring conditions change, etc. An analysis of any and all observations may be a necessary part of this step so that the radio 'knows where it is at'. In Figure 5.4 the trigger is a new assignment of spectrum to the radio.
2. Once the decision-making process is triggered the high-level action of 'shape transmitted signal' has to be *decomposed* into a set of manageable decisions. In Figure 5.4 two broad areas that need decisions are shown: waveform type and antenna usage. There are undoubtedly

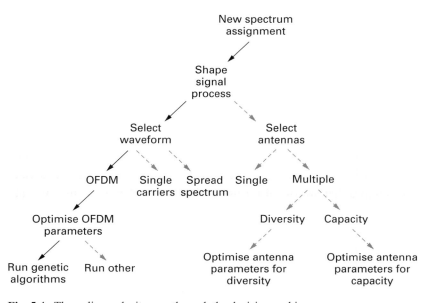

Fig. 5.4. The radio works its way through the decision-making process.

many more but these suffice for illustration. Following the waveform type selection trace, we see this is further decomposed into a choice between OFDM, single carrier or spread spectrum waveforms. Then following the OFDM trace this reduces to an optimisation problem involving selecting OFDM parameters such as FFT size, modulation technique, cyclic prefix details, pilot details, etc.

3. Once the decision or set of decisions that needs to be made has been identified, it is then necessary to decide how the decision or decisions will be made. This involves deciding what, if any, optimisation techniques should be used. Part of this process may involve looking at the kind of decisions that need to be made. Recall that the taxonomy of decision types that was introduced in unilateral and multilateral centralised decisions will use different approaches from distributed multilateral decisions. There may be also time issues, accuracy issues, search space size issues, that will push towards one approach rather than the other. The phrase meta-decisions is often used to describe these decisions about decisions. In Figure 5.4 the meta-decision is to use genetic algorithms again for purposes of illustration.

4. The particular decision-making technique or optimisation technique must then be *initialised* and set in motion. Depending on the technique it will need to be fed different kinds of information. All techniques will need to be seeded with a definition for goodness of fit or 'best' and many will need to be initialised in a wider variety of ways. In Figure 5.4 the genetic algorithm would be initialised with the population of chromosomes that starts the process. Constraints need to be taken into account and fed into the decision process.

5. Finally the output of the process is put into action. The output may first of all be tested to ensure it is an allowable action or the optimisation process itself may be carried out in such a way as to ensure that unallowable actions are not selected.

As mentioned already but worth stressing again, observations play a major role in the process in Figure 5.4. Remember that the term 'observation' is used in this book in a very broad sense and can mean any inputs relating to the radio environment, the physical environment, the regulatory environment, or observations of user preferences and behaviours.

The observations can be used at the outset to understand the context in which the radio finds itself and hence trigger a decision-making process in the first place.

Overall we can say that *knowledge* has been brought to bear on interpreting observations to trigger actions and influence selections. Knowledge has been applied to the process of decomposing the high-level action into manageable and meaningful decisions (X different independent and Y dependent decisions are needed). Knowledge has been applied to help make meta-decisions (optimisation technique A is good for situation X and optimisation technique B is good for situation Y). Knowledge has been applied to the initialisation of the selected optimisation processes (I will only get a good answer if I run technique A for Z seconds and if I use population of genes W).

To support this kind of functionality, a cognitive radio has to *capture* existing wisdom and knowledge through some formal way of *representing* the knowledge. Whatever the representation used, the cognitive radio needs to be able to *reason*[9] about the knowledge it has and deduce next steps or make conclusions. Hence knowledge representation and reasoning are vital to the decision-making process.

5.3.1 A little about knowledge representation and reasoning

Knowledge representation is the study of how knowledge about the world can be represented and what kinds of reasoning can be done with that knowledge. It is a large field and the short discussion here merely aims to capture some simple concepts. The purpose of representing knowledge is to make it *machine readable*. There are different ways for representing knowledge. And there are different methods for reasoning about that knowledge. The kind of reasoning that can be applied is often related to how the knowledge is represented in the first place.

Methods of representation range from propriety approaches that can only be understood and used by the creators of the representation to

---

9 To reason is to think in a connected, sensible or logical manner; to employ the faculty of reason in forming conclusions (in general, or in a particular instance).

very structured systems that allow for sharing of knowledge more easily. Typically the more structured and formal the representation, the easier it should be for different systems and entities to use the same knowledge base. It is worth listing a few examples of knowledge representation approaches. Knowledge can be represented using a simple database in which machine-readable data is stored. Knowledge can be represented using *production rules*, sometimes called IF-THEN rules. They can take various forms such as IF condition THEN action or IF premise THEN conclusion. This is also termed *rule-based* knowledge representation. Knowledge can be represented using an *ontology*. An ontology provides a shared vocabulary, which can be used to model a domain, that is, the type of objects and/or concepts that exist within that domain, and their properties and relations. Ontologies provide a way of specifying types and type hierarchies as well as relationships and relationship hierarchies. Several ontology languages have been developed in recent years. For example, the Web Ontology Language (OWL) is a family of knowledge representation languages for authoring ontologies.

Typically some kind of *inference engine* makes sense of rule-based systems. While each rule is an independent item of knowledge, the inference engine allows for the chaining together of rules to infer other knowledge. Techniques called *forward and backward chaining* of rules are used to string rules together and make inferences. The downside of rule-based systems is that they are limited to fixed capabilities designed into their rule set and cannot easily be extended. They are obviously dependent, like many of the knowledge representation techniques, on the expertise of those who create them. Inference is also used to reason with ontologies. Very expressive ontology languages allow ontologists to specify first-order logic constraints between terms and more detailed relationships such as disjoint classes, inverse relationships, part-whole relationships, etc. Hence the more expressive the ontology, the more sophisticated the kind of inferences that can be carried out.

It is also possible to *reason by analogy*. This involves the transferring of knowledge from one analogous situation to another. *Case-based reasoning* (CBR) is a well-known kind of analogy making. It is the process of solving new problems based on the solutions of similar past problems.

In case-based reasoning a database of existing cases is used to draw conclusions about new cases. The reasoning used here invokes procedures like *pattern matching* and various statistical techniques which can be used to find an existing case that is similar to the case that is being explored. This kind of reasoning therefore is very different to the kind of inference process used in rule-based approaches and with ontologies. Case-based reasoning tends to be successful when there is a rich database of cases. However, it is easy to add new cases as they become available.

A cognitive radio could, for example, use a database of different propagation environments which also include lists of radio configurations suited to each environment. A database of observations could be constructed. A set of rules could be composed to take account of the processes laid out in Figure 5.4 and to specify how decisions are broken down into subdecisions. Alternatively a cognitive radio ontology could be designed. The idea of a cognitive radio ontology has garnered some interest in recent years. It does seem attractive to create an ontology that can be widely understood in the cognitive radio community. Such an ontology could encompass representation of the cognitive radio object itself as well as the environment in which the cognitive radio finds itself. To create a useful ontology all terms of relevance would have to be identified, classes of objects and hierarchies of objects would need to be specified.

5.3.2 Summary part 2

The main point from Section 5.3 is that the core optimisation processes need a lot of extra support functionality. In essence a cognitive radio needs to represent, organise, store and analyse knowledge so that appropriate optimisation routines can be evoked. The discussion here merely gives a flavour of what is possible. Each of the knowledge representations discussed here and the many more that exist capture different aspects of knowledge and suit different situations. In the first instance it makes sense that some uniform mechanism for knowledge representation be selected for use in cognitive radio. Currently cognitive radio ontologies seem to be the favourite option. Any suitable form of reasoning could then be used by individual cognitive radios. This is an area where standards bodies can play a role. However, to get really advanced cognition, it is likely that different knowledge representations will be needed. A quotation from Marvin Minksy [3] captures this:

> ... in the 1960s and 1970s, students frequently asked, 'Which kind of representation is best?' and I usually replied that we'd need more research. ... But now I would reply: To solve really hard problems, we'll have to use several different representations. This is because each particular kind of data structure has its own virtues and deficiencies, and none by itself would seem adequate for all the different functions involved with what we call common sense.

5.4 Taking regulations into account when making decisions

A very important issue in cognitive radio is how to take regulatory policies into account when making decisions. Regulatory policies, like any other knowledge, need to be represented in a machine-readable manner and the cognitive radio needs to reason about the polices and make decisions that are in keeping with these policies. Clearly this topic could have been treated under the general heading of knowledge representation and reasoning. It was in fact alluded to in the previous section when we spoke about observations, including inputs about regulations, feeding into the decision process and possibly constraining choices of actions. However, it is dealt with here separately, not just because of its importance but because there is some emerging consensus on how to deal with this topic and that consensus is around *policy-based management.*

Policy-based management is an administrative approach that is used to simplify the management of a given endeavour by establishing policies to deal with situations that are likely to occur. The policy simply defines *what* actions are to be taken in the given situation but does not specify *how* the action is to be carried out. There are different policy-based management approaches. Here we will look at the policy framework defined

by the IETF/DMTF.[10] The IETF/DMTF policy framework consists of
four basic elements:

1. The policy management tool: An administrator uses the policy management tool to define the different policies that are to be enforced in
   the network. In the case of cognitive radio, that administrator could
   be a regulator or could be any other administrator that had jurisdiction
   over a given area or entity.
2. The policy repository: This stores all the policies that have been
   generated using the policy management tool.
3. The policy decision point (PDP): The policy decision point is an entity
   that is responsible for retrieving the policies from the repository, and
   for interpreting them and deciding on which set of policies ought to
   be enforced by the PEP.
4. The policy enforcement point (PEP): PEPs are logical entities that can
   take actions to enforce the policies.

The policy rules can take any machine-readable form. They can, for
example, take the form of rules and have the 'if condition, then action'
structure. The condition can define the circumstances whether that be
time, location, user, or more detailed and specific conditions about the
environment. A simple example for a cognitive radio could be 'If in
Ireland do not use frequency range X.' There is also much talk of creating
policy languages around cognitive radio ontologies.

In a policy-based management system, the policy decision point and the
policy enforcement point need not be co-located with the policy repository. Typically these would be in the cognitive radio. The policy repository
would be located somewhere globally accessible. The cognitive radio
would use the policy rules, which it fetches via the policy decision point,
to guide/constrain its actions. The policy enforcement point corresponds
to the point in the system where the configurations of the radio are put into

10 The Internet Engineering Task Force (IETF) is the body that defines standard Internet
   operating protocols such as TCP/IP. The Desktop Management Task Force (DMTF) is an
   industry consortium of more than 120 vendors, established in 1992 and committed to
   making PCs and servers easier to understand, use, configure and manage.

action. The simplification in management is obtained primarily by centralising the definition of policies in a single repository, and by defining abstractions that provide a machine-independent specification of policies. The policy rules can be set by a regulator or some controlling entity, and cognitive radios can download rules as they move from one jurisdiction to another. The centralised policy repository can be changed over time to reflect changing policies. This allows rules to be very dynamic and open up new possibilities for spectrum management. This will be further addressed in Chapter 8.

There is no reason why policy-based management could not be used for more general knowledge representation and reasoning as well as for regulatory policies. Other knowledge can be stored as policies. And there can be both local and global repositories. For example, there can be policies set by a user (e.g. 'never spend more than X euro on communication per week without notifying the user'). There can be policies about cognitive radio behaviours that have purely technical concerns. It is possible to establish hierarchies of policies so that regulator-related policies can override local policies should conflicts arise.

Policy-based management is emerging as a very popular approach in cognitive radio and there a number of examples in the research world at least in which policy-based management is used to control cognitive radios.

5.5 The decision-making process: part 3

We have now seen that optimisation forms a core part of the decision-making process but that, in order to run optimisation routines to select the appropriate actions, reasoning is needed to make sense of what is happening in the radio and to decide what happens next. To reason we need to represent knowledge in some machine-readable way. However, there is more still to consider. Reasoning deduces and infers conclusions from knowledge that exists. To add to existing knowledge a cognitive radio must *learn*. Before looking at some techniques which can be used to learn it is useful to examine how learning can help or is needed in a cognitive radio.

5.5.1 Learning that is additive

Learning can be of great benefit to a cognitive radio in improving its overall performance. This learning can happen in simple or more complex ways. On the more simple side of learning a cognitive radio can *memorise* actions it takes in given scenarios, *recognise* when a similar scenario arises and reuse the actions, rather than evoke a new optimisation process to set the action parameters. It can memorise and learn the preferences of users and take preemptive action to address user needs.

At a more complex level a cognitive radio can apply advanced techniques to spot patterns and events hidden among observation data, not obvious to a human but deducible using machine learning methods. We can go back to the world of spectrum sensing to show how this might help. Consider here again the case of a primary and secondary user. The secondary user must detect available spectrum but more importantly if it uses the spectrum it must continue to sense in order to detect the return of the primary user in a timely fashion so that it can vacate the band. A cognitive radio that manages to learn the behavioural patterns of a primary user could use this extra information to improve its performance. So, for example, it may be the case that at certain times of day or in certain locations the white spaces that emerge tend to be more long lived. Or there may be predictable qualities to the behaviour of primary users. This kind of learning can emerge from longer-term analysis of spectrum sensing data and can be used by the radio in different ways. White spaces that have a higher probability of having longer durations can be selected over others. This has the potential to greatly reduce or even eliminate interference to primary users. Alternatively the learning can be used so that the sensing process can be tailored to zone in on frequency ranges that will be more likely to contain white spaces, etc. This can cut down on computation times and increase the likelihood of success. Knowing something about primary user patterns can also help when a secondary user has to vacate a frequency and move to another because some kind of advanced planning as to where to go next may be possible.

A cognitive network can also use learning to help with coexistence issues. For example, two neighbouring cognitive networks could, through observation over time, learn about each other's frequency and spatial footprints and use this information to self-adjust in order to coexist. Cooperation between neighbouring networks can exist today based on manual intervention, and the automation of this process through the use of cognitive networks is a logical step for the future. It is probably useful at this stage to point out that a cognitive radio can learn from itself as well as from others. As stressed many times in this book, a cognitive radio is simply a node in a network, and gains can be made through the learning of others. Automated coexistence management for neighbouring cognitive networks is an example of where many nodes would feed into the learning process.

A cognitive radio can also bring learning to bear on the decision process itself. The cognitive radio can learn to make better meta-decisions – it can learn to choose different optimisation techniques or to initialise the chosen techniques more appropriately. Of course to do this the radio will have to rate its performance against some benchmark and take note of any improvements which might occur. Making more accurate or faster decisions can help improve the overall performance of the radio.

5.5.2 Learning that is crucial

So far we see that learning provides a means of improving performance which is always welcome but not necessarily critical. However, certain improvements in performance may lead to widespread acceptance of cognitive radio and in this context the learning would be a crucial factor. The spectrum sensing example illustrates this well. One of the key issues for dynamic spectrum access networks is to avoid causing interference to existing users. If the potential for meaningful interference is removed because of a greater ability to predict primary user behaviour then learning in this case may be the key to opening the door to serious deployment of cognitive radios. However, aside from these scenarios, there are some actions which cannot be taken without the involvement of learning techniques and this is the brief focus of the following paragraphs.

In our first look at the decision-making process we noted that the optimisation process can be defined as the process involved in selecting the 'best' choice from the list of available choices in order to reach some kind of goal or at least get as near as possible to the goal. If you can define 'best' then the process is somewhat straightforward, even in the case of large numbers of choices. Defining 'best' typically involves defining the relationship between choices (actions) and the goal (objective function) for the given set of circumstances (as revealed via observations). There are many situations in cognitive radio in which this relationship cannot be precisely specified and the cognitive radio will have to learn what it, in fact, is. It can learn by trial and error or in other ways. In fact, without some kind of learning, there is no way in which scenarios like this can be accommodated. In other words, learning is a crucial part of the system. As we saw in Chapter 3 there are so many possible 'knobs' that can be set on a cognitive radio that it is likely that many scenarios will arise in which the exact relationship between the possible actions and the objective or goal of the decision-making process will not be known. The other main area in which learning is crucial is in dealing with unforeseen or unplanned events. Again some kind of learning mechanisms that facilitate a trial-and-error approach may prove useful.

5.5.3 A taste of learning

Machine learning is the broad name sometimes given to the whole range of techniques that support learning. This is again a very large field of study and there is a wide range of mechanisms that can be used to learn. The aim, as before, is to give a flavour of what is available.

There are three broad learning modes, namely supervised learning, unsupervised learning and reinforcement learning. *Supervised learning* is a machine learning technique for learning from training data provided by a knowledgable external supervisor. In *unsupervised learning* there is no training data. In this case the aim is to discover inherent structure in the data with only inputs and no outputs. *Reinforcement learning* is learning what to do – how to map situations to actions – so as to maximise a numerical reward signal. The learner is not told which actions to take, but instead must discover which actions yield the most reward by trying them. In the most interesting and challenging cases, actions may affect not only the immediate reward but also the next situation and, through that, all subsequent rewards. These two characteristics – trial-and-error search and delayed reward – are the two most important distinguishing features of reinforcement learning. To obtain a lot of reward, a reinforcement learning agent must prefer actions that it has tried in the past and found to be effective in producing reward. But to discover such actions, it has to try actions that it has not selected before. Reinforcement learning is particularly relevant for the scenarios described in Section 5.5.2 in which it is not at all clear what actions should be taken to reach a desired goal.

There are a large number of different supervised, unsupervised and reinforcement learning techniques in existence. From the cognitive radio it is clear that in cases where there are examples of actions and the circumstances in which those actions are taken, then supervised learning techniques can be used. And obviously, in cases where no such input (circumstances or scenario details) and output (actions) pairs exist, unsupervised techniques or reinforcement learning is needed. Using the input/output vocabulary, supervised learning predicts outputs for novel inputs based on example pairs (input, output). The task of unsupervised learning is to predict outputs based on novel inputs, without example but knowing what kind of actions may feature in the output list. And the task of reinforcement learning is to predict outputs for novel inputs without a clue as to what actions should feature in the output list. Some machine learning techniques can be executed in supervised, unsupervised and reinforcement modes, while others are suitable for use in one mode only.

Before looking at different machine learning techniques which call on one or all of the learning modes just introduced, it is worth making a few comments about learning versus reasoning. To a certain extent there is an artificial categorisation of techniques in this chapter. Many of the reasoning techniques are part of learning processes and it is just a case of slightly furthering the process to encompass learning. A good example is case-based reasoning. In circumstances in which the case under exploration does not match any existing cases, it seems logical for it to be added as a new category of cases. This is in fact learning. This is why some texts treat reasoning and learning together. However, the separation is useful when it comes to mapping out what functionality is needed for a given implementation of a cognitive radio. A flavour of some common techniques follows.

Pattern recognition is a well-known learning technique. Given a pattern, its recognition/classification may consist of one of the following two tasks: supervised classification in which the input pattern is identified as a member of a predefined class, or unsupervised classification in which the pattern is assigned to a hitherto unknown class. Note that the recognition problem here is a classification or categorisation task, where the classes are either defined by the system designer (in supervised classification) or are learned based on the similarity of patterns (in unsupervised classification). The supervised version here is similar to the case-based reasoning as per the comments in the previous paragraph. Pattern recognition is an example of a technique that would play a great role in the spectrum sensing example of Section 5.5.1. The radio in this case would use the techniques to discover patterns in the behaviour of the primary users, based on spectrum availability measurements taken over time.

Pattern recognition can be used in all of the three areas to which learning applies in cognitive radio, outlined above. It can be used to recognise certain combinations of observations and corresponding actions (supervised), improve understanding of observations and form new classifications not noted by humans (unsupervised), to learn what actions should go with patterns of observations, especially when exact relationships between the observations and actions are not understood. Two examples of pattern recognition approaches are *template matching* and *statistical classification*. Template matching needs training data and is the simplest of the three approaches mentioned here. In the statistical approach, each pattern is represented in terms of d features in d-dimensional space. The goal is to choose those features that allow pattern vectors belonging to different categories to occupy compact and disjoint regions in a d-dimensional feature space. New and existing patterns can be created.

Neural networks could in fact be validly labelled as a pattern recognition technique, as it can be difficult to categorise and separate different learning techniques. An artificial neural network is an information processing paradigm that is inspired by the way biological nervous systems, such as the brain, process information. The key element of this paradigm is the structure of the information processing system. It consists of an interconnected group of artificial neurons that work in unison to solve specific problems. Artificial neural networks learn by example. Learning involves adjusting the synaptic connections that exist between the neurons until the correct output is obtained. An artificial neuron is a device with many inputs and one output. A system can be modelled by a large number of neurons. This approach is particularly useful when it is not clear what combinations of actions can lead to a desired output or behaviour as described in Section 5.5.2.

5.5.4 Summary part 3

There are many learning techniques ranging from very simple memorisation techniques to more complex approaches. The field of artificial intelligence (AI) and machine learning is a large one with plenty of overlap in the cognitive radio domain and offering plenty of opportunities. Incorporating learning in a cognitive radio is very possible. There are a number of research papers, such as [4] and [5], describing work that has been carried out using different learning techniques, which show how feasible it is. Just as in the case of using optimisation techniques, it is more a matter of selecting learning mechanisms that suit the time scales of the radio's day-to-day life. Learning techniques have generally been targeted at single sample problems rather than embedded in a wide range of cognitive radio functionality but this reflects the stage where the field is currently, rather than indicating that it will not be possible to use learning in a more widespread fashion. A final key point is that learning is crucial when dealing with unknowns or unplanned scenarios.

5.6 Conclusions

The various elements of the chapter come together in the *cognitive engine*. The cognitive engine is the name given to the *brain* of the cognitive radio, the part where all the optimisation, reasoning and learning 'stuff' happens, the part that takes in observations and spits out decisions. The cognitive engine can also include all that is associated with dealing with regulatory policies. Sometimes this functionality is encompassed in a policy engine. There are numerous and many very elaborate diagrams of what a cognitive engine looks like but essentially they all boil down to varying mixes of elements that can optimise, reason and learn combined with some kind of knowledge base.

From the discussion in the chapter so far it may seem as if a very extensive optimisation process would need to be run in order to identify which kinds of knowledge representation make sense, which optimisation algorithms should be used, which reasoning mechanisms are needed and what kind of learning techniques should be applied, because there are so many choices available. It is important to keep the situation in perspective though. An all-singing, all-dancing cognitive radio will need to use a number of different knowledge representations as indicated in Marvin Minsky's comment, a range of optimisation techniques, a selection of reasoning methods and a number of different machine learning approaches. However, many of the applications discussed in Chapter 1 require no such extreme levels of functional sophistication. Enormous amounts of flexibility and indeed high levels of future-proofing can be gained with each additional small increase in functionality. The real and essential challenge is how to map out what should be included and how far the various techniques can go.

Figure 5.5 attempts to capture the notion of there being vast amounts of choice in the implementation of a cognitive engine. It may be that a few

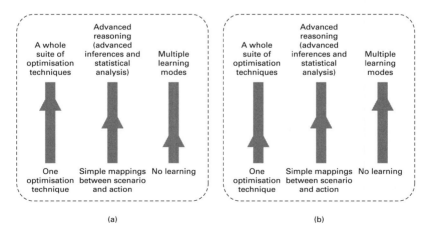

Fig. 5.5. Some options for the functionality of a cognitive radio.

optimisation techniques combined with strong learning abilities is best (option b). Or it may be that a large suite of optimisation techniques combined with good reasoning capabilities and very little learning satisfies all the needs (option a) or indeed some other combination of elements. Much of this may be driven by the applications. Albert Einstein said 'theories should be as simple as possible, but no simpler'. This can be translated into the cognitive radio world as 'cognitive engines should be as simple as possible, but no simpler'.

References

1. T. Weise, *Global Optimization Algorithms: Theory and Application, 2008*. Downloadable at http:www.it-weise.de/projects/book.pdf
2. D.H. Friend, R.W. Thomas, A.B. Mackenzie and L. DaSilva, Distributed learning and reasoning in cognitive networks: Methods and design decisions, in *Cognitive Networks, Towards Self-Aware Networks*, ed. Q.M. Mahmoud. Chichester: Wiley, 2007.
3. M. Minsky, Logical vs. analogical or symbolic vs. connectionist or neat vs. scruffy, *Al Magazine*, **12**:2 (1991), 34–51.
4. C. Clancy, J. Hecker, E. Stuntebeck and T. O'Shea, Applications of machine learning to cognitive radio networks, *Wireless Communications, IEEE*, **14**:4 (2007), 47–52.
5. T.W. Rondeau, Application of artificial intelligence to wireless communications, PhD Thesis, Virginia Polytechnic Institute and State University, 2007.

6   Security in cognitive radio

6.1 Introduction

The chapters so far have looked at the main functionality of a cognitive radio through an exploration of the 'observe, decide, act' cycle. We now step back from this and look at security issues specifically related to cognitive radio. This chapter is the shortest chapter of the book. This is not indicative of the level of importance of the topic of security and cognitive radio but of the fact that cognitive radio security has to date received much less attention than other topics.

All communication systems need to be made secure to operate. Typically any users of a system have to authenticate themselves on the network. *Authentication* is the process of determining whether someone or something is, in fact, who or what it is declared to be. Some authentication processes may involve the simple use of a password but others are more complex. The use of digital certificates, issued and verified by what is known as a *Certificate Authority* (CA) as part of a *public key infrastructure,*[1] is an example of a more stringent process. Following authentication, *authorisation* processes, to ensure data and services are accessible only to those who have the correct entitlements, are needed. During all communication *privacy* may need to be guaranteed. *Encryption* is typically used to achieve this either using a public key

---

1 In cryptography, a public key infrastructure (PKI) is an arrangement that binds public keys with respective user identities by means of a Certificate Authority (CA). The user identity must be unique for each CA. The binding is established through the registration and issuance process, which, depending on the level of assurance the binding has, may be carried out by software at a CA, or under human supervision. The primary function of a PKI is to allow the distribution and use of public keys and certificates with security and integrity. The specific security functions in which a PKI can provide foundation are confidentiality, integrity, non-repudiation and authentication.

infrastructure, a symmetric cryptographic approach[2] or hash algorithms.[3] Eavesdropping by man-in-the-middle attacks[4] must be avoided. Today there are many privacy issues arising from digital traces we leave behind as we work our way through the network electronic world, and more sophisticated ways of using aliases and multiple identities are emerging for dealing with this. For systems where money must exchange hands for services, secure accounting is needed to properly trace the transactions and subsequently settle bills. *Non-repudiation* techniques are needed as part of the accounting process. And in all networks it is important to secure against denial-of-service attacks[5] and watch out for single points of failure.[6]

These basic issues are security issues for any network. Over the years a wide range of authorisation, authentication, encryption, accounting, non-repudiation and other techniques have been developed for wireless networks. More complex versions of these techniques have been designed for distributed networks as the distributed nature of a system can often add to the challenge. However, an exception to this is that distributed networks may have no obvious single points of failure. A cognitive network, like any other wireless network, will have to adopt many of these security measures. We are interested in identifying the security issues which are *unique* to cognitive radio and cognitive networks. Hence the chapter is organised around various features of cognitive radio which may raise security issues.

---

2 Symmetric cryptography is based on the use of a secret key that is shared between the communicating parties, i.e. the same key is used to encrypt and decrypt the data. Typical algorithms used are DES, RC4 and RC5.

3 Hash algorithms are used to calculate a message digest, which is a value computed by a one-way function (hash function) from some data. A one-way function means that it is impossible to construct the original piece of information from the digest and that it is extremely difficult to construct another piece of information that has the same digest. Typical algorithms used include MD5 and SHA-1.

4 In cryptography there is a form of active eavesdropping in which the attacker makes independent connections with the victims and relays messages between them, making them believe that they are talking directly to each other.

5 A denial-of-service attack (DoS attack) or distributed denial-of-service attack (DDoS attack) is an attempt to make a computer resource unavailable to its intended users.

6 A single part of the network, which if it fails will result in the unavailability of the entire network.

6.2 The strength and weakness of being able to observe

To examine cognitive radio from a security perspective we again return to its definition. One of the strongest features of a cognitive radio is the fact it takes in and processes so much information. In other words it observes much. It then adapts its behaviour on the basis of the observations through making appropriate decisions and taking action. We have seen that the observations include an understanding of the environment in which the radio operates, an understanding of the communication requirements of the user(s), an understanding of the regulatory policies which apply to it and an understanding of its own capabilities. We have learned that cognitive radio is 'all about understanding context'.

While other radios use information to trigger adaptive behaviour the breadth of information that is used in a cognitive radio is much more vast. Some information is particularly crucial. For example, information about the regulatory policies that apply is vital in ensuring a radio behaves correctly. Information about the radio environment is essential if interference is to be properly managed, while information about user requirements might be of lesser or greater importance. Whatever the importance though, all action springs from the information the radio gets. Hence if the observations are wrong then pretty much everything else will be wrong too! In other words, the very strength of a cognitive radio, i.e. its ability to be aware of its surroundings and circumstances through the gathering of information and observations in order to decide how to act, is also a huge weakness.

The dependency of a cognitive radio on observations opens up a plethora of opportunities for security attacks. Very simply put, if false data can be generated or if data can be tampered with, the behaviour of the cognitive radio can be controlled.

6.2.1 Physical fakes

Attackers can generate signals that are correctly observed by cognitive radios but are in fact only generated to 'confuse' the cognitive radio. This type of attack makes sense especially in the context of dynamic spectrum access networks, in which the cognitive radio senses the presence of an incumbent or primary user. The malicious terminal emits signals that emulate the characteristics of the incumbent. This malicious terminal can cause the cognitive radio to think that there is no unoccupied spectrum. This will either result in the cognitive radio not transmitting at all or moving unnecessarily to another frequency. Whether this is a likely event or not remains to be seen. The attack takes some effort and equipment to stage. And in one sense, the more agile the radio, the more resistant it is to such an attack as it will be able to use a plethora of other frequencies (unless in the unlikely event the attacker occupies all frequencies). Should this kind of attack turn out to be a strong possibility, it is possible to create systems that are robust to this. Cognitive radios may have to ensure that any feature-based sensing uses incumbent/primary characteristics that are non-forgeable in detecting their presence.

This kind of issue tends to arise only when considering cognitive radios that need to share spectrum with existing users who are not to be disturbed or put upon to incorporate new features to facilitate their cognitive radio neighbours. In many applications in which systems are designed anew, security features including authentication processes can be built into the system. Policing of spectrum usage should also help identify rogue transmitters, although this is made very difficult if the transmitter is on infrequently.

6.2.2 Physical degrading of performance to stop sensing

Attackers can raise the noise floor in the vicinity of a cognitive radio to make it impossible to detect signals that should normally be detected. On the one hand this seems like a problem for any radio system, i.e. it can be jammed. However, the issue here is that it is just the ability of the cognitive radio to sense that needs to be disrupted. This will take much less effort than the jamming of its ability to communicate. The problem with these jamming attacks is that it may be very difficult to distinguish between malicious intent and a badly behaving cognitive neighbour. For example, if a neighbouring cognitive node ends up unintentionally clipping the signal it is transmitting, and as a result causes alarming amounts of spectral regrowth around its transmission profile, it can cause problems to other cognitive radios. Spectrum policing/monitoring may be needed to ensure that there are no malicious or misbehaving nodes. And, a point we will return to later, if it is possible for the cognitive radio to rely on multiple sources of information from other parties, there may be a means of getting the information it needs when it cannot generate it itself.

6.2.3 Physical tampering with data

Attackers can physically tamper with stored data such as policy databases. Any tampering with the regulatory policies that are used by the radio is of grave concern. It is, however, perhaps safe to assume that any policy database would be well secured. There also may be ways of ensuring content integrity. Content integrity means that the receiving party can be sure that the transferred information is exactly what the other party originally sent. Integrity can be established by computing a message checksum or digest[7] and attaching it in every message before the message is sent. The receiving party can then re-compute the digest and compare it with the original digest. The cognitive radios can also authenticate policies before enforcing them, to ensure they came from the correct party. Reiterating the point that was made at the outset of this section, in radios in which data plays such a great role in understanding context and in deciding actions, all data needs to be treated with care and to be secure.

6.3 The double-sided coin of collaboration

Collaboration among cognitive nodes is a feature of many cognitive radio applications. Collaboration can take place in the context of sensing, in the context of any self-organising process, in the context of sharing resources, for bargaining purposes, during normal operation (especially in distributed scenarios in which nodes forward information for each other) and in many other instances. Collaboration is an enabling feature

---

7 A message digest is a number which is created algorithmically from a file and represents that file uniquely. If the file changes, the message digest will change.

of a cognitive network. And again, while other more traditional networks involve elements of collaboration, the cognitive network seems to go further.

The problem with relying on or using collaboration extensively is that a malicious or misbehaving node in a collaborative network can cause damage. It may even be possible for a few well-placed malicious nodes to cause a disproportionately large amount of damage. In fact this topic would have been equally relevant in the previous section on observations, especially as collaborative sensing is key in many cognitive radio applications: for example it can be crucial in dealing with hidden node problems. A misbehaving or malicious radio can purposely not report the fact it has sensed an incumbent, leading to spectrum being detected as unoccupied when in fact it is occupied. It can purposely insert all sorts of other false information into the system – for example, it can report false neighbour numbers and give the impression that the network is more sparse that it actually is. It can give false accounts of traffic flows and congestion, and can encourage the use of unsuitable routes.

On the one hand we can solve this issue with security measures. Strong authentication of all users and careful admission control policies might prevent such an event happening. Content integrity measures, such as those discussed earlier in the chapter, can also be applied. And some kind of reputation system may prove useful. A reputation system is a type of collaborative filtering algorithm which attempts to determine ratings for a collection of entities, given a collection of opinions that those entities hold about each other. For example, eBay has a simple reputation system rating the participants, which despite its simplicity seems to work.

The reputation system leads back to a more important point. Security attacks and malicious behaviour aside, a cognitive network will typically often be heterogeneous in nature. Different nodes will have different abilities to sense. And, despite every good intention, a node may insert false or poor data into the system. To deal with this, there needs to be careful design of all collaborative protocols. For example, thresholds can be set that more or less say 'unless X neighbours say they saw Y, then Y did not happen' and these thresholds can be dynamic. This threshold example is very simple but there are more sophisticated filtering and information fusion options and ways of weighting observations. The idea of weighting observations also suggests the use of a reputation system. Hence the design of robust collaboration mechanisms for cognitive networks should be good enough to force the attacker to have to compromise a large number of nodes in order to make any impact.

6.4 Physically tampering with the cognitive radio

Attackers can tamper with the cognitive radio to make the radio behave badly. For example, if the radio were to be compromised in such a manner as to override any regulatory policies, the consequences would be stark. The fact that many cognitive radios, as will become clearer in Chapter 7, use much software makes this more likely, especially in circumstances where that software may be downloaded over the air interface. In this case the kinds of viruses we see from downloading software on to PCs may manifest themselves, as may viruses and attacks targeted to cognitive radios. Many of the techniques mentioned already such as authenticating where the software comes from and using some form of data integrity checks can help here. In the design of the software aspects of the cognitive radio itself, it may be necessary to ensure that a modular approach is adopted to ensure that if one part of the system is compromised other parts remain safe.

6.5 The single points of failure

Any system may have a single point of failure. Cognitive radio networks may have some particularly worrying ones, however. The first which springs to mind is the cognitive radio control channel or the cognitive radio beacon. While it is by no means clear that a control channel for cognitive radio, in any of the discussed application areas, will exist, it is nonetheless important to discuss the control channel as a very obvious single point of failure.

A control channel can be attacked by jamming the channel physically or congesting the channel with useless information. Protocols can be put in place to stop the latter, though the former is more of a problem. Beacons can be compromised by disrupting timing. In some beaconing systems, the beacons are transmitted at certain times and the cognitive radios must listen for their presence in a particular time pattern. This possibility of attack of a control channel or beacon is one of the strong arguments against such things. However, the answer may be to think of a control channel in a different way. A particular insight from Mackenzie[8] is to use the Internet as a control channel. A cognitive radio could use all and every means available to connect to the Internet to receive bootstrapping instructions. Hence there is no one frequency or physical channel that is used to get the information needed.

There are other single points of failure that may arise in a cognitive network, depending on the application in use. Applications which use spectrum brokers to manage spectrum and other resources are one such example. Again there is an obvious point of attack but, as in the case of a policy database, tight security would need to be maintained.

6.6 Application demands

The last point focusing on unique security issues in cognitive radio is not in fact unique! The point is that many cognitive radio applications are ones in which security is paramount. Two examples spring to mind. The first is public safety and the second is real-time spectrum trading.

While all public safety applications require high levels of security, cognitive radio may open up some more issues. Currently many public safety groups cannot easily communicate with each other. There tends to be fragmented use of different technologies and different frequencies, making it difficult for the fire service to talk to the ambulance service to talk to the police service but at the same time affording some kind of security if only because of so many differing systems. Cognitive radio is seen as a means of seamlessly bringing all the services together which will without a doubt call for all the security measures mentioned in this chapter and more to be used. There may be additional need for security

8 Made by Allen B. Mackenzie during a workshop in Dublin in 2008.

measures to support such functions as group formation. Chief fire fighters and chief of police may want to be in a private subgroup of the greater public safety operation and within this subgroup there may be a need to have certain hierarchies in place to control communication flow. Hence security techniques for this kind of dynamic formation of groups is another type of security that might be needed.

Spectrum trading applications will call on the transfer of monies and the use of some kind of auction mechanism. Just like public safety, this kind of application has a great need for secure transactions. In some of the just-in-time systems, micropayments may be useful. The main point again is to emphasise that cognitive radio applications can often make high security demands.

6.7 An example of security in action

As we have been saying throughout the book, cognitive radio is still in a very formative phase. Hence there are limited examples of any working systems. In order to draw on an example related to security we return again to the IEEE 802.22 standard. The IEEE 802.22 is a standard for delivering wireless broadband in TV white spaces. The standard takes account of security issues through the specification of *security sublayers*. The security sublayers focus on integrity of the data, identification of users, association of valid identities to users, device authentication, authorisation processes, confidentiality, privacy protection of data from eavesdropping and non-repudiation prevention among other things. While the security aspects are not fully formed it is worth listing some of the directions and recommendations that are being considered:

- The aim is to use database functions for authentication, authorisation, key management, service suspend/resume, relocation, anti-cloning, etc.
- There is a requirement that firmware be tamper-proof to prevent unauthorised modification to firmware and/or functionalities. Any attempt to load unapproved firmware into a device must render it inoperable.
- A CPE must authenticate itself with the network every time it registers with the network. This prevents unauthorised CPEs from entering the

network and prevents unauthorised basestations from emulating an authorised basestation.

- There is a stipulation that passwords and keying information must now be passed 'in the clear' through the air interface.
- There is an interest in the use of credentials. Each subscriber has a set of credentials that describe what the subscriber is 'allowed' to do. A standard set of credentials must be developed for this purpose.
- Suitable cryptographic algorithms must be employed. There is a recommendation that facilities should be defined in the protocol for the use of alternative cryptographic algorithms in varying localities, which can replace algorithms as they are obsoleted.
- Message integrity should be facilitated to ensure the accuracy of delivered data and that the data has not been altered in an unauthorised manner.

These few examples do not give any details of how any of these measures will be implemented but they do reflect some of the issues discussed in the chapter. And they show that security is being considered as an integral part of the first potential commercial cognitive radio standard.

6.8 The silver lining

The bulk of the chapter has focused on security issues that are unique to cognitive radio. However, the potential benefits of cognitive radio should not be neglected when discussing the topic of security, lest the landscape look completely bleak. In Chapter 5, the decision-making process was described in detail. The impression one gets on leaving that chapter is of a highly sophisticated device that is capable of complex reasoning, advanced pattern recognition and deep learning. While it was emphasised that not all cognitive radios will have all of these features, it is nonetheless true that cognitive radios will have good processing capabilities. It is likely that some of the cognitive radio functionality, e.g. classification/data mining and learning techniques, can be put to good use in the delivery of advanced security measures.

A second point worth noting is that, within the life span of a cognitive radio, it will run many applications that will require it to update its policy database, download new pieces of software, interface with a database, communicate with others. While in previous sections of the chapter we looked at the downside of this behaviour, there is also an upside in that the radio can be kept aware of the most up-to-date security concerns.

6.9 Conclusions

There are many security challenges in wireless networks and there are extra security challenges that are unique to cognitive radio. However, there are ways of making cognitive radios secure enough to operate with confidence and to require significant effort to compromise. Like any system, there are tradeoffs when security is involved. The processing of information can be slowed because of the many checks and, given that cognitive radio depends so much on the processing of observations, there are challenges to secure in as lightweight a manner as possible. Whatever the challenges, the main point for cognitive radio should be to learn from the mistakes of others. The most significant mistake to make is to not build security into the system from the beginning. Conventional wisdom very much says that 'adding at the end never works'.

ative ratio will almost all of the time function as a node in a network

7   Cognitive radio platforms

7.1 Introduction

In Chapter 1 the working definition for cognitive radio used throughout this book was presented. That definition ended with the statement 'A cognitive radio is made from software and hardware components that can facilitate the wide variety of different configurations it needs to communicate.' In this chapter we look at the hardware involved. There is no one right way to build a cognitive radio so the chapter merely aims to give a sense of what kind of hardware can be used and some of the related performance issues.

7.2 A complete cognitive radio system

In a cognitive radio receiver, the antenna captures the incoming signal. The signal is fed to the RF circuitry and is filtered and amplified and possibly downconverted to a lower frequency. The signal is converted to digital format and further manipulation occurs in the digital domain. On the transmit side the opposite occurs. The signal is prepared and processed and at some stage is converted from digital to analogue format for transmission, upconverted to the correct frequencies and launched on to the airwaves via the antenna.

Throughout this book we have been using the terms 'cognitive radio' and 'cognitive node' interchangeably. The reason for this is that a cognitive radio will almost all of the time function as a node in a network. Therefore it is useful to think of the complete cognitive radio system in terms of a communication stack. The physical (PHY) layer and the media access (MAC) layer have already been defined. Network layers, security layers, application layers may sit on top of the PHY and MAC. How high up the communication stack 'cognition is brought' is a design choice.

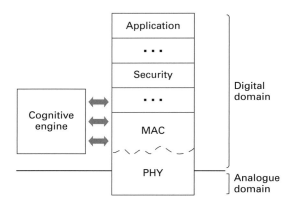

Fig. 7.1. The complete cognitive radio system perspective.

The cognitive radio must contain some kind of cognitive engine that makes decisions and directs operations. The mechanisms for making the cognitive decisions could be implicitly built in to various layers of the communication stack. Using this approach the PHY layer would contain elements of decision-making and learning, as would other layers. It is more useful, for the moment, to think of the cognitive engine as residing in the digital domain but as some kind of separate entity that interfaces with the other elements of the system and has a cross-layer vantage. Figure 7.1 attempts to capture this view in some kind of unified diagram.

Figure 7.1 is a very high-level overview and an infinite number of real implementations could be mapped to that diagram. There are a number of points to note. Firstly the diagram emphasises that the PHY layer cuts across the analogue and digital domains. It is a matter of design choice as to where the line actually is. Secondly the expected neat line between the PHY layer and the MAC layer has been removed. The reason for the broken and curved line is to signal the breakdown of the strict interface between the traditional PHY and MAC layer that is very much a feature of cognitive radio. This is because the MAC layer has a major role to play in the sensing process and as a result there is a stronger coupling of PHY and MAC layer functionality. In fact there are those who maintain that cognitive radio needs a complete redesign of the communication

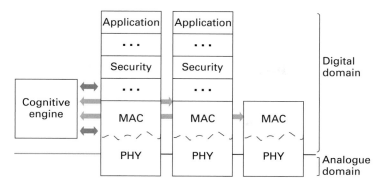

Fig. 7.2. An example of a system with multiple RF frontends.

stack and who would see Figure 7.1 as outdated and believe a whole new paradigm is needed. However, the figure is useful in conveying the functionality of the digital aspects of the radio. The exact design of the system that delivers this functionality is an open question.

A cognitive radio can have more than one RF frontend. There may be separate circuitry for sensing or even perhaps for some kind of dedicated control channel. That aside, there is much interest now in multiple radio (transceiver) systems. This is not unlike a modern mobile phone with wireless LAN and Bluetooth transceivers as well as GSM and 3G radios. However, in the case of cognitive radio there would be N similar radios, all open for configuration, so that multiple network topologies can exist at different frequencies.[1] Figure 7.2 expands Figure 7.1 to take account of this. Note that the extra radios can be supported by full or partial stacks.

Figure 7.2 and Figure 7.1 contain very high-level diagrams. There is possibly an infinite number of ways in which the cognitive radio(s) in these figures can be implemented. There are multiple ways to design the hardware and the software components of the system. And there are enormous choices around what functionality should be included. All that we can hope to achieve in this chapter is to give some insight into a few

1 This is a concept which is of core interest in the WNaN (Wireless Network after Next) project. The WNaN project can be found at http://www.darpa.mil/sto/solicitations/WNaN/

key concerns that must be taken into account in the design of a cognitive radio. Hence we focus on the issues that arise in selecting suitable digital hardware as well as the performance of the analogue aspects of the radio and the impact they have on the radio as a whole.

7.3 Cognitive radio platforms: digital hardware

Figure 7.1 can be used to reflect on the kind of tasks that need to be carried out in the digital domain of the radio. At the PHY layer there is the need for many heavy-duty signal processing tasks; at the MAC layer there is much need for tasks which organise data flow and can be very dependent on getting timings correct. Much of the sensing functionality of the radio is directed and controlled from here, for example. At higher layers various protocols exist, many of which need to organise the sending and receiving of packets, perform analysis of data, maintain caches, etc. The security layer may call for mathematical tasks involving encryption. The cognitive engine may use many different kinds of decision-making algorithm, optimisation techniques, AI techniques, etc. The policy engine may do much reasoning and deduction as well as have database-related jobs. A designer will be faced with the job of specifying all the tasks that are needed, deciding how they are to be implemented, defining how the different components of the system interface with each other and mapping the design to a hardware platform.[2] Certain design approaches will suit certain digital platforms. Hence a key question for the designer is which platform to use.

7.3.1 Basic digital hardware options

There are a number of standard digital hardware options that can be used in a cognitive radio. In choosing a platform, we are typically interested in three features, namely *flexibility*, *performance* and *power consumption*.

---

2 Any one implementation of a cognitive radio may be geared towards a specific application and hence have no requirement for one or more of the kinds of processing task listed here, but for now we will proceed with the broad picture in mind.

Flexibility is a key factor in cognitive radio as radios must reconfigure themselves in order to suit prevailing conditions or user demands, etc. Different kinds of cognitive radio and different designs may call for different levels of flexibility. Performance relates to speed. Some tasks, such as heavy-duty signal processing, for example, may require exceptionally high speeds for execution and may not be implementable on slower platforms. Power consumption is very important in handheld devices but is becoming increasingly important in terms of sustainability and the environment. In the following sections six different hardware platforms for digital processing are briefly described. The descriptions are followed by an analysis of the options from a flexibility, performance and power consumption perspective, which gives a good indication of the tradeoffs faced by the designer.

ASIC

An ASIC is an application specific integrated circuit. An ASIC, as the name suggests, is a circuit designed for a specific application, as opposed to a general purpose circuit, such as a microprocessor. An ASIC is highly optimised for the task at hand and hence can deliver good performance. It cannot be reconfigured. ASICs are low in cost, have high performance and low power consumption. There are many ASICs in mobile phones.

FPGA

A field-programmable gate array (FPGA) is a semiconductor device consisting of a logic layer and a configuration layer. The logic layer is an array of basic logic blocks, such as AND and XOR, or more complex logic functions like multipliers or memories. The configuration layer is a grid of programmable interconnects of the logic elements, which is used to implement a given algorithm. The designer implements a given algorithm by deciding how the blocks are connected. FPGAs enable a high degree of parallelism, i.e. all used logic blocks execute at the same time, rather than sequentially. Hence they offer very good performance. The phrase 'field-programmable' highlights the fact that the FPGA's function is defined by a user's program, i.e. is programmed in the field, rather than by the manufacturer of the device. Code can be loaded on to the FPGA's configuration memory to create the desired system. Traditionally the FPGA configuration is loaded from an external persistent memory (e.g. a PROM) at system startup and stays there for the full system lifecycle. Recent FPGAs allow dynamic partial reconfiguration, i.e. parts of the FPGA can be reconfigured while the remaining parts continue executing. Thus, functionality can be loaded on-demand, significantly increasing the device's flexibility and efficiency. However, the cost of flexibility is higher power consumption than the ASIC. FPGAs are often used in systems when new ideas are prototyped. They are used in basestations. WiMAX designs that use FPGAs are available.

ASIP

ASIP stands for application specific instruction-set processors. ASIPs attempt to take some of the good characteristics of ASICs (such as low power and low cost) and add some flexibility and limited levels of programmability such as that offered by digital signal processors (DSPs), microprocessors and microcontrollers. ASIP combines a software programmable processor and application specific hardware components. Hence there is some flexibility in the system. The hardware components can be used a number of times in an instruction pipeline scheme. The specific hardware components make sure good speeds can be achieved. However, ASIPs are more power hungry than FPGAs.

DSP

A digital signal processor is a microprocessor whose architecture is specially designed for numerical computations on discrete number sequences specifically tailored to the processing of signals. A DSP is a programmable device, with its own native instruction code. This helps in the writing of code to implement the algorithms of interest. Once written the code can then be downloaded to the DSP chip. Though most general-purpose microprocessors and operating systems can execute DSP algorithms successfully, the specialised digital signal processor provides more optimised performance. The DSP core typically consists of an arithmetic logic unit, accumulator(s), multiple and accumulate (MAC) unit(s) and data and address busses. The multiply and accumulate operation ($a \leftarrow a + b * c$)

is very common in DSP algorithms. The fact that DSPs cater for this explicitly makes them very suited to signal processing applications. Very many digital signal processing functions are needed in the PHY layer of a cognitive radio, making DSP chips attractive platforms. DSPs consume more power than FPGAs but tend not to be as fast. DSPs can be found in mobile phones.

GPU

A graphics processing unit or GPU is a dedicated graphics rendering device. GPUs exist in many PCs and game consoles. Modern GPUs are very efficient at manipulating and displaying computer graphics, and their highly parallel structure makes them more effective than general-purpose CPUs for a range of complex algorithms. A GPU can sit on top of a video card, or it can be integrated directly into the motherboard.

GPP

A general-purpose processor is a family of microprocessors and micro-controllers best suited for performing a broad array of tasks that are not specifically tailored for any particular application. All PCs contain general-purpose processors. Programs can be very easily written in a wide range of programming languages, to run on a GPP. Hence GPPs are very flexible, though they consume much more power than other options.

7.3.2 Making choices

Figure 7.3 compares the hardware options from the perspectives of the three key features of flexibility, performance and power consumption. The figure is based on one originally generated by Noll [1]. A line demarcating the platforms that follow a software paradigm and those that follow a hardware paradigm has been added, as well as details about GPUs. The hardware/software paradigm line simply emphasises the design mindset that is used in the different platforms.

As is clear from the figure, the different options all have different advantages and disadvantages. ASICs score high for low power and very high performance. What is not obvious from the figure is that they also

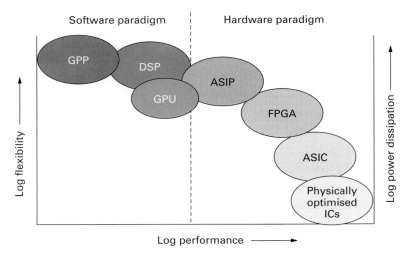

Fig. 7.3. Different digital hardware options from a flexibility, performance and power consumption perspective.

have low production costs. Their main disadvantage, especially in the context of cognitive radio, is lack of flexibility. Any updates would require a whole redesign. Hence if ASICs were to be used in an adaptive system, all eventualities would have to be foreseen and an ASIC for each designed. FPGAs are usually slower than their ASIC counterparts and draw more power but are faster than GPPs due to the fact they support parallel designs. FPGAs provide for reconfigurability, partial and complete, and this makes them an attractive platform for cognitive radio. They are fast and because of their parallel structure they can cope with difficult intensive computations and deal with large amounts of data. While they are more expensive than ASICs they are still less than all the other options. ASIPs come next in Figure 7.3. They are more flexible than FPGAs and ASICs while still retaining some of the more low-power characteristics.

DSPs move towards higher power consumption, are slower than ASIPs, FPGAs and ASICs but on the other hand are very flexible. As distinct from many GPP solutions, they deliver hard real-time performance, which means that you always know how long a task will take. The fact that standard software programming languages can be used to programme DSPs makes them very attractive. However, despite the option of using high-level languages such as C, many designers work at the assembly language level. This approach tends to give more control over the exact implementation of the algorithm and more optimised behaviour. As can be seen, GPPs score for flexibility. However, they use more power and tend to be slower than other options. The power consumption is a big negative. On the plus side there is reduced product development time and cost because of the availability of large sets of development and debug tools. In terms of performance Figure 7.3 does not tell the whole story as high-performance GPPs are becoming more DSP like with optimised instructions. However, on the downside high-performance GPPs timing behaviour is stochastic and can only support soft real-time performance, and the GPP is the most costly of all the options.

The fact that each of the different hardware options has different advantages and disadvantages is well recognised. Take for example what is happening in the FPGA world. The FPGAs that are currently in use contain much more than basic programmable logic. The family of FPGAs from Xilinx, known as Virtex-4, is a good example of this. When you buy an FPGA from the Virtex-4 family you get up to 200 000 logic blocks depending on the exact FPGA chosen. Some of the Virtex-4 FPGAs contain PowerPC processors, a general purpose processor. The FPGA also contains digital signal processors. So, even though Xilinx still use the term FPGA to describe their product, it is in fact a combination of different processors. The reason for this is the fact that Xilinx, while still having a major focus on the FPGA logic blocks, recognises that many current applications need many different kinds of processor in order to get the levels of functionality they require. The need for different hardware in cognitive radio solutions is echoed in this description from Louis Bélanger from Lyrtech [2]:

The received RF data is either down converted to an intermediate frequency band using super heterodyne or directly converted to baseband using direct conversion radios.[3] Converting IF to baseband in a super

---

3 These are two of the most common types of radio architectures that can be found. In the case of the superheterodyne receiver the system is designed to receive a given range of frequencies and convert these incoming frequencies to what is termed an intermediate frequency, IF, before the signal is digitised and sent onwards for further process. In direct conversion there is no intermediate stage, as the name would apply.

heterodyne architecture with IF digitisation involves a digital down conversion process. This can be realised either in a DSP with a very low IF (generally under 1 MHz) or in dedicated ASIC in the case of single waveform-based radios for best in class cost and power consumption. But in multi-mode systems supporting multiple waveforms, this requires flexible logic and involves the use of FPGAs. The dynamic real-time processing of DSP can then take over from the FPGA in the demodulation of the signal. The process is repeated in reverse on the transmit path.

The system of error control for data transmission in wireless modems, called forward error correction, can be implemented either in DSP or logic gates depending on the type of encoding/decoding algorithms used. For example, the Reed-Solomon encoding and decoding algorithms along with encoding for others like convolutional and turbo codes is easy and better to implement in DSP for cost/power benefits. However, the more complex cycle-intensive technique of the convolutional or Turbo error correction algorithms are best implemented using hard logic gates integrated in a processor or with the use of a reprogrammable FPGA. In that sense, the FPGA plays a very strong co-processing role in multi-mode systems allowing for flexibility in the support of multiple protocols on the same radio.

The extracted data packets from the physical layer (layer 1) of the modem are passed on to the media access control layer (layer 2 or MAC) for management of the physical connection to the network (also referred to as network processing). MAC layer processing involves encoding and decoding of packets into bits, transmission to and from network interface as well as flow and conflict management of data packets within a channel. This networking processing requires the efficiency of Real Time Operating Systems (RTOS) and involves a number of control processing functions. The best processing elements to implement MAC functions as well as the memory management needed in RTOS require microprocessors or micro-controllers generally referred to as General Purpose Processors (GPPs). Thus the combination of GPP, DSP, and FPGA components is a necessity in multi-mode systems to implement multi-band multi-protocol radios.

The long description, copied verbatim, emphasises that it is not so much about choosing one piece of hardware over the other, but about using multiple pieces of hardware and choosing which task is executed where.

7.3.3 Multi-core systems

The discussion, thus far, has defined the basic digital hardware options. We now look at *multi-core* systems. These systems are becoming increasing relevant in the design of cognitive radios. A multi-core system consists of, not surprisingly, multiple processing cores. These cores may be similar (a homogeneous architecture) or of differing types (a heterogeneous architecture). They may consist of two up to hundreds or thousands of processing cores. Irrespective of the number, the term multi-core applies. As stated in the PicoChip white paper on Practical Programmable Multi-Core DSP,[4] 'the real world is parallel . . . Desktop computers have to fool their users into thinking that several tasks are executing in parallel when this is usually not the case: the computer is in fact working sequentially, but very quickly.' Multi-core systems allow for parallel implementations of programs that bring with them a performance increase without a corresponding power consumption increase.

The multi-core architecture uses some kind of network or bus infrastructure to allow the cores to communicate. Via this infrastructure the cores can share data, access instructions, and signal between themselves for control purposes. The moving of data has become one of the bottlenecks in performance for DSPs and GPPs. To address this, multi-core systems make use of both local and global memory within the overall system. Well-designed multi-core architectures allow data stores to be distributed throughout the system, in whatever way makes most sense for the application. In multi-core architectures the communications fabric can also substitute for memory accesses by allowing direct communication between the processing elements.

Homogeneous multi-core GPP

A multi-core GPP is an integrated circuit to which two or more GPP processors have been attached, for enhanced performance, reduced power consumption and more efficient simultaneous processing of multiple tasks. The use of multiple GPPs or multi-core processors is becoming

---

4 This very accessible and well-written paper can be downloaded from www.picochip.com.

more widespread. Very many PCs are now shipped with dual-core or quad-core processors. (A dual-core processor contains two cores and a quad-core processor contains four cores.) A dual core, for example, is comparable to having multiple, separate processors installed in the same computer. However, because the two processors are actually plugged into the same socket, the connection between them is faster. A two-fold performance gain over a single-core processor is the ideal. In practice, performance gains are said to be about 50%. In the case where multiple processes are running, the operating system will ensure that different processes are placed on different cores and the advantages of the system will be gained. However, if the aim is to use one process on the multi-core system, it needs to be designed to ensure that it can take advantage of the multiple cores.

There are a number of different paradigms for taking advantage of mulit-core. It is possible to use *task parallelism*. In this, a given task is split into subtasks, and the subtasks are spread across the cores and run in parallel. In this case the different tasks work on the same data. An alternative approach is to use a *data parallelism*. In its simplest form this involves using the same operations on all the elements of a collection of data, such as an array or set. To use these approaches the code has to be appropriately designed. To move existing sequential programs to multi-core platforms will call for a redesign.

7.3.4 The cell processor

The cell processor is another platform that contains multiple parallel processors. The complete name of the cell processor is the cell broadband engine. The cell was developed with game and multimedia applications in mind. The Sony Playstation 3 contains a cell processor and is a well-known example of a very affordable version. The cell architecture contains multiple processors known as synergistic processor elements (SPEs) and a power processor element (PPE) which controls what is happening on the SPEs. The cell also contains an on-chip memory controller, and a controller for a configurable I/O interface. Figure 7.4 shows a broad outline of its architecture. Like the other multi-core systems, the

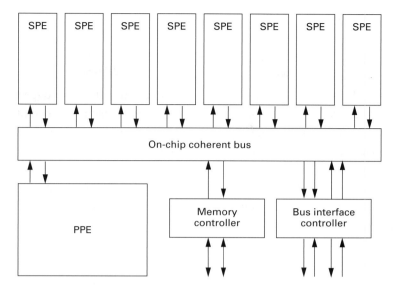

Fig. 7.4. The cell architecture.

power of the cell again lies in its ability to parallelise operations and hugely increase speed. Typically control code runs on the PPE. This is programmed exactly the same way as a desktop processor. The programming of the SPEs can be carried out, as in the case of the multi-core GPP, using task or data parallelism approaches. Communication between SPEs is usually done via a shared memory but additional communication methods exist as well. Mailboxes and signals are methods of sending smaller messages or signals between different parts of the chip. If you want an SPE to wait for an event you can set an SPE to wait for a mailbox or signal; this will be useful for real-time processing. The cell is emerging as a very powerful platform for cognitive radio applications. Even though it is not as flexible as a GPP, it is still very flexible in comparison to many other platforms while delivering high performance.

Multi-core DSP

As in the case of GPPs, multi-core DSPs also exist. Initially coprocessor technology was used to boost performance of DSPs. A coprocessor offloads specialised processing operations and hence takes the burden off the DSP core. The coprocessors often run at the same frequency as the DSP core, therefore 'almost double' the performance for targeted applications. Multi-core DSPs boost performance further by facilitating the parallelisation of tasks as discussed above. While single-core DSPs are levelling off at around 1 GHz performance, multi-core DSPs operating at lower frequencies are reaching effective performance levels of 2 GHz and more. A four-core 500 MHz device results in 2 GHz of performance with lower power per MHz than single-core DSPs. It is possible to utilise a multi-core device in different ways, such as a DSP farm where the cores operate independently. The interfaces route data to and from each core with no interaction between cores. In other cases the cores work together to complete a task.

The picoArray hardware from PicoChip is an example of a multi-core DSP architecture. The picoArray is a heterogeneous array of hundreds of processing and storage elements, optimised specifically for signal processing tasks. The cores include memory, multiply and accumulate engines, and general-purpose 16-bit processors. The cores can be interconnected, as defined by the system architect, at software compilation time. New designs can be easily downloaded and executed, making picoArray ideal for use in reconfigurable systems. In the picoArray each process gets a fixed resource on which to run, and a fixed communications resource to serve it. This eliminates the need for dynamically scheduling processes, and fixed resources produce deterministic behaviour. There are three different types of processor, all with the same basic architecture but different memory size or coprocessors. The three kinds are STAN (standard, for DSP datapath operations), MEM (with more memory, used for local control or memory intensive tasks like buffers or look-up tables) and CTRL (with largest memory for control tasks). The picoArray provides a good example of the trend towards multi-core architectures.

Highly heterogeneous multi-cores
Finally we look at a multi-core system that is highly heterogeneous. Zhang and colleagues in the University of Twente have designed heterogeneous tiled architecture, where tiles can be various processing elements including general-purpose processors, FPGAs, ASICS and

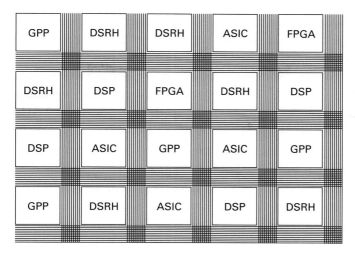

Fig. 7.5. A heterogeneous multiprocessor architecture.

domain-specific reconfigurable hardware (DSRH) modules. Figure 7.5 is a conceptual diagram of the heterogeneous multi-core system. In their paper on the architecture, Zhang *et al.* [3] point out:

The tiles in the System-on-a-Chip (SoC) are interconnected by a Network-on-Chip (NoC). Both the SoC and NoC can be dynamically reconfigurable, which means that the programs (running on the reconfigurable processing elements) as well as the communication links between the processing elements are configured at run-time. Different processing elements are used for different purposes. The general purpose processors are fully programmable to perform different computational tasks, but they are not energy efficient. The dedicated ASICs are optimised for power and cost. However, they can not be reconfigured to adapt to new applications. FPGAs which are reconfigurable by nature, are good at performing bit-level operations but not that efficient for word level DSP operations. The Domain Specific Reconfigurable Hardware (DSRH) is a relatively new type of processing element, where the configurable hardware is tailored towards a specific application domain.

This echoes the discussion in Section 7.3.2. Different hardware is needed for different tasks. Here, however, one chip contains it all rather than in the Virtex-4 example where discrete hardware elements are combined. Initial mechanisms for mapping applications to this type of architecture have been designed, though the architecture is still in the research stage. It does indicate the future of multi-core and the many possibilities that exist.

7.4 Cognitive radio platforms: the analogue part

We now turn to look at the analogue part of the cognitive radio. The field of RF design is very large and complex, and in this book we cannot hope to focus on the design of radio transceivers. The purpose of this section is therefore more limited and has two objectives: to get a sense of the basic elements in a radio transceiver, and to understand how to rate transceiver performance. This is needed for a designer as well as a user of RF transceivers. As a designer, performance metrics can be set as targets. As a user, performance metrics are indicative of the limits of the system. A designer of cognitive radios may aim for certain performance levels. A functioning cognitive radio needs to understand its own capabilities and limitations to decide on actions.

7.4.1 A general look at the RF frontend

The quotation in Section 7.3.2 mentioned two receiver architectures, namely a superheterodyne and a direct conversion receiver. Rather than select a specific receiver or indeed transceiver architecture we instead focus on a more generic transceiver architecture that captures the essential elements of the radio. We base the generic architecture on that used in Chapter 2 of Jeffrey Reed's book on *Software Radio* [4].

After the signal is received by the antenna some filtering takes place. Then a local oscillator and a mixer come into play. Essentially a reference signal generated in the local oscillator is mixed with the incoming signal so that some signal at a lower, more manageable frequency can be generated (though some architectures will not have such stages). More filtering occurs, followed by automatic gain control before the signal gets converted to digital format. On the transmit side the signal is converted to analogue format, upconverted to the relevant frequency range, filtered,

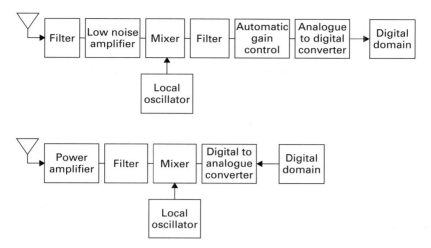

Fig. 7.6. This generic transceiver architecture will be used as a guide in this chapter.

amplified and transmitted via the antenna. Note that there is only one antenna depicted in Figure 7.6. There may in fact be multiple antennas in use, as is the case for the MIMO systems mentioned in Chapter 3. The main point to be taken from Figure 7.6 is that all RF transceivers consist of some combination of amplifiers, filters, mixers, local oscillators, gain controllers and antennas. Figure 7.6 gives a flavour of the kind of combination that arises.

What is important now is to understand broad performance issues rather than details of individual radio components. To do this we need to look at ways of describing the radio behaviour or performance.

7.4.2 Radio performance metrics

To get a sense of radio performance we start by looking at the receiver and at how well it can deal with interference and noise. To do this we need to focus on a range of key parameters that is used to describe receiver performance.[5]

---

5 Some of the discussion that follows was influenced by the following website:
  http://www.radio-electronics.com/info/receivers/index.php, accessed on 30 June, 2008

One of the most important performance metrics of a receiver is its *dynamic range*. The dynamic range of a radio receiver is essentially the range of signal levels over which it can operate. The low end of the range is governed by its *sensitivity*, while the high end is governed by its overload. In this book we look on dynamic range as central to an understanding of the radio.

We start with the limits on the low end of the dynamic range. The sensitivity determines the weakest signal a radio can detect. The main limiting factor in any radio receiver is the noise generated. For most applications either the *signal-to-noise ratio* or the *noise figure* is used to describe noise. If the radio were perfect then no noise would be added to the signal as it passed through the radio. The signal-to-noise ratio would be the same at the output of the radio as at the input. However, each element of the RF chain in the receiver, such as those shown in Figure 7.6, adds noise. A noise figure can be given for each individual element in the RF chain or for the whole chain. The noise figure is the ratio of the signal-to-noise ratio at the input to the RF chain (or element) and the signal-to-noise ratio at the output of the chain (or element). It is always the noise performance of the first element of the receiver that is most crucial. At the frontend the signal levels are at their lowest and even very small amounts of noise can be comparable to them. At later stages in the radio receiver the signal will have been amplified and will be much larger. The same levels of noise as are present in the first elements of the RF chain will be a much smaller proportion of the signal and will not have the same effect. The term *noise floor* is sometimes used and can be defined as the measure of the signal created from the sum of all the noise sources and unwanted signals within a system.

At the high end of the dynamic range a number of issues come into play. Overload performance governs how well the radio can receive strong signals. To understand this, we start by discussing the non-linear behaviour of the amplifier in the RF chain. In an ideal world the output of an RF amplifier would be proportional to the input for all signal levels. However, beyond a certain level, an amplifier enters a non-linear range and the output and the input of the amplifier are no longer directly proportional to each other. In the non-linear region the signal starts to become

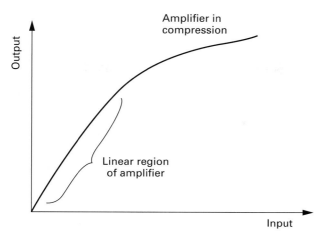

Fig. 7.7. Non-linear behaviour.

compressed resulting in the characteristic shown in Figure 7.7. Automatic gain control (AGC) can be used to reduce the level of the input signals and thereby make sure that the radio does not enter the non-linear region. However, it cannot always prevent the frontend stages from being overloaded.

When in the non-linear region bad effects occur that hamper the performance of the receiver. We touched on this in Chapter 3 in the discussion on interference. The term *intermodulation distortion* (IMD) was introduced. This distortion occurs when two signals suitably spaced in frequency combine in such a manner so as to generate unwanted signals at new frequencies. These new signals, or *intermodulation products* as they are called, are named after the kind of frequencies at which they occur. There are second-order, third-order and higher-order intermodulation products. Figure 7.8 shows some of these intermodulation products generated from two frequencies, F_1 and F_2. The third-order products are the ones that cause problems as they land near the signals of interest. The second-order products can be filtered out.

An understanding of third-order intermodulation products is needed to understand the concept of *third-order intercept point*, IP3, which is an important measure of performance. The IP3 is a theoretical point and

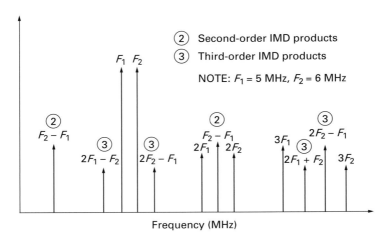

Fig. 7.8. Some intermodulation products.

is not a measured quantity. It can be understood as a measure of the linearity of a device. As signal strength is increased by 1 dB the third-order intermodulation products increase by 3 dB. The IP3 point is the hypothetical output signal level, at which the third-order tones would reach the same amplitude level as the input tones. When this point is reached, the mixer theoretically becomes saturated and a further increase in input signal will not cause a further increase in output volume. In other words the signal enters the compressed region shown in Figure 7.7. Hence the IP3 point is the point that gives us some kind of understanding of what limits the higher levels of the dynamic range.

Some other descriptors and metrics also have an indirect bearing on the dynamic range. One of these is *selectivity*. The selectivity of a device determines whether the receiver is able to pick out the wanted signal from all the other ones around it. Good selectivity can reduce unwanted out-of-band signals and is very much based on how well the filters perform. A filter has a response which determines how much signal gets through and how much does not. A typical response is shown in Figure 7.9. Good filters have a steep roll-off, i.e. the transition from the passband to the stopband is steep. The 'Q' of a filter is a way of specifying this. This is a measure of the sharpness of the amplitude response. It is

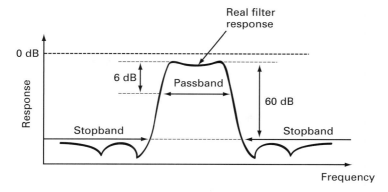

Fig. 7.9. The response of a filter. The passband covers the frequencies that are admitted. The stopband shows the frequencies that are blocked.

in fact defined by $Q = f_c/(f_{low} - f_{high})$ where f_{low} and f_{high} are the frequencies at the -3 dB response points. High Q filters are sharp filters and quite selective. This relates to the ability to block adjacent channel interference.

When a very strong out-of-band signal appears at the input to a receiver it is often found that the sensitivity is reduced due to what is termed *blocking*. The block specification is a measure of the relative strength of the interfering signals to the desired carrier frequency that cause the receiver sensitivity to degrade by 3dB. The effect of blocking is that the frontend amplifiers run into compression as a result of the interfering signal – the dynamic range reduces if the sensitivity reduces. The further away the unwanted signal, the more it will be reduced by the frontend tuning and the less the effect will be.

Figure 7.10 attempts to capture the various concepts we have discussed and relate them to the central concept of dynamic range.

Our discussion of an RF transceiver has really just been about the RF receiver. The transmitter will have a noise figure as well and a similar discussion can be had about the filters it contains and their ability to shape the transmitted signal. The main concern is the ability of a transmitter to generate a nice clean signal. The other main concern in a transmitter relates to power levels. An entity called peak to average power ratio

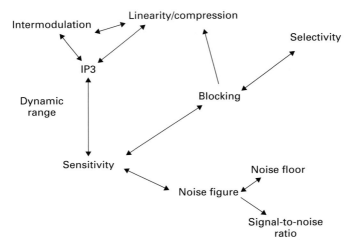

Fig. 7.10. One view of the confusing world of performance metrics for radios.

(PAPR) comes into play. This gives an indication of the difference in levels between average and peak power. However, this is very much a function of the kind of waveform that is being transmitted. The reason why this is of interest relates to the receiver, in that transmitted signals with high PAPRs can force a receiver into regions of non-linear operation.

Making choices

When it comes to cognitive radio, it is helpful to think of transceiver performance in two ways.

1. In the first instance we can say transceivers must be designed to meet the needs of the cognitive radio application. A quick look at Chapter 1 will help recall the type of applications that are foreseen for cognitive radio. The exact demands on the RF frontend can be application dependent.
2. In the second instance we can think of a cognitive radio having whatever transceiver it has, and using its 'cognitive engine/brain/smarts' to best use its attributes.

It is worth giving some examples to get a sense of the kind of demands that exist in case 1. In dynamic spectrum access applications there is a demand that the RF frontend be frequency agile over a wide spectrum of operation. This may call for tunable components. Tunable filters, for example, can be used to zone in on different frequency bands as the RF frontend is reconfigured to hop from frequency to frequency. There may be demands for good selectivity and very high Q filters to make sure the radio transmits carefully within its assigned spectrum as well as to ensure it carefully picks out the signals intended for it. The need for wideband or tunable components goes right through the radio system. Amplifiers need to be wideband or tunable as well. Antennas may need to be very wideband also. However, we will return to antennas later.

Other cognitive radio applications may call for the supporting of multiple different kinds of standards, in a software radio-like manner. This means that different waveforms, signal bandwidths, modulation formats, signal levels and blocking specifications may need to be considered. As a guide using requirements of existing systems, cellular standards have low to medium bandwidths, very high dynamic range requirements and difficult blocker environments. Wireless local area network standards mean lower power levels are involved, less dynamic range is needed and there are fewer blocker considerations. They have, however, high signal bandwidths and high-order modulation requirements. This means higher signal-to-noise ratios and hence better noise performance in the receiver.

Sensing requirements will place other demands on the cognitive radio. The sensing receiver may need a high dynamic range to be capable of picking up both high and low signals. The radio may need to be made of high-quality low-noise components to ensure low noise figure receivers. Even though signal processing techniques can be brought to bear and longer observation times used, there will nonetheless be some limitations. Energy detection techniques, as was pointed out in Chapter 4, have signal-to-noise ratio walls.

Software control of the analogue frontend is possible, i.e. it is possible to move the operating characteristics of a transceiver in real time by software commands. This may be the approach needed in applications that require very varying types of performance.

In case 2, in which we use the cognitive radio to drive the RF frontend in as efficient a manner as possible, the various performance metrics may feed into the decision-making processes. Hence, if the radio has a limited dynamic range, the radio will not take actions that mean it ends up communicating in an overly adverse environment. The radio may choose not to communicate using waveforms with high PAPR, so does not run as much risk of operating in its non-linear region. This would mean avoiding waveforms like OFDM or alternatively using extra signal processing techniques such as clipping to reduce PAPR.

7.4.3 The antenna

There can be many demands made on the antenna of a cognitive radio. The need for wideband antennas was already mentioned briefly. Cognitive radios in fact may draw heavily on all sorts of antennas, ranging from single-element antennas to arrays of antennas. As a radio cannot function without its antenna, it is worth spending some time discussing antenna issues for cognitive radio.

In order to understand some of the requirements for the antenna, a few basic facts are needed. First the size of antenna has some relationship with the frequencies of the signal being transmitted. The phrase 'half-wave dipole', for example, refers to an antenna that is roughly half the wavelength of the transmitted frequency. Radio waves propagate with the speed of light, c.[6] The relationship between frequency and wavelength is $c = f\lambda$. Hence lower frequencies mean large antenna sizes. While antenna lengths are typically a half or quarter the length of the wavelength, there are a whole host of techniques in existence, mainly involving the use of special materials, that allow the antenna size to be further reduced. However, you cannot get away from the fact that at certain frequencies antennas will be large.

The second point to note relates to the definition of the bandwidth of the antenna. Antennas have a gain. The gain is the amount the antenna

6 The speed of light is commonly set at 3×10^{-8}m/s.

adds to the transmitted signal.[7] Typically an antenna has a frequency range over which the gain will be maintained – or at least over which good gain is maintained. This tells you the bandwidth of the antenna. In recent years there has been research targeted towards making very wideband antennas, particularly because of UWB communication systems.[8] For example, some UWB antennas have to cover the 3.1 and 10.6 GHz band. A particularly successful UWB antenna is the *antipodal Vivaldi antenna* which can operate over that bandwidth and more and be less than 6 cm × 6 cm × 0.7 cm in size. Typical wideband antenna families in use are planar and non-planar monopole antennas, TEM-horn antennas[9] and dipoles.

In Chapter 3 beamforming was discussed as a means of manipulating the spatial footprint of the transmitted or received signals – basically to control interference. This kind of spatial filtering is achieved by using an array of antennas rather than one antenna and giving different weights (emphasis) to the array outputs. The antenna weights can be either fixed in advance or calculated in an adaptive fashion according to various optimisation criteria such as those discussed in Chapter 3. Very complex antenna systems may suit basestations more than mobile handsets. In Chapter 3 MIMO techniques for increasing capacity were also discussed. MIMO also necessitates multiple antennas at the transmitter and the receiver. The IEEE 802.11 are looking at MIMO techniques for wireless LAN. Hence the idea of using multiple antennas on portable devices will become much more the norm in the future. Polarisation diversity uses the horizontal and vertical polarisation of electromagnetic waves to theoretically double the antenna diversity order without the need for

---

7 Because there are lots of different antennas, it is necessary to define a reference gain which can be used as a means of comparison. An isotropic antenna is one in which the transmitted power is radiated equally in all directions. This antenna is commonly used as a reference. Gains which are quoted relative to this isotropic antenna are quoted in dBi, where i denotes isotropic.

8 Recall that UWB was described briefly in Chapter 3.

9 The basic TEM-horn antenna has been designed for radiation of short pulses with duration of about 1 ns. It consists of a pair of triangular conductors forming a V structure in which a transverse electromagnetic (TEM) wave propagates along the axis of the V structure.

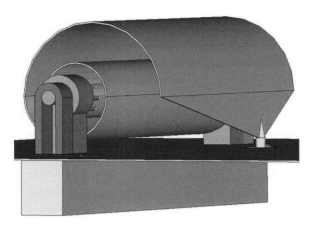

Fig. 7.11. A reconfigurable antenna. The frequency of operation is changed by physically curling up the antenna more or less tightly.

adding new antenna elements. Chapter 6 of Reed's book on *Software Radio* [4] gives a good overview of these kinds of antenna under the general heading of smart antennas.

All of the antennas discussed so far are designed for operation over a given frequency range. In the case of very wideband antennas the aim is to make the frequency range large enough to cover the frequencies of interest. An alternative approach is to use a reconfigurable antenna whose physical properties can somehow be changed in order to change the operating (or resonant) frequency of the antenna. Looking towards the future, antenna solutions have been identified that might be reconfigured dynamically to adjust the resonant frequency, gain and polarisation of the antenna. In Figure 7.11 the frequency of the antenna is actually changed by causing the antenna to curl up more or less tightly. The curling process alters the physical parameters of the antenna in such a manner as to change the operating frequency. This antenna was designed by Ruvio *et al.* [5]. In this case the antenna is suitable for use from 2.2 to 15 GHz and its radiation patterns show good omnidirectionality over that range. The use of multiple antennas and reconfigurable antennas does require more complex circuitry and puts extra burden on the cognitive radio. However, the gains that can be made in exploiting these techniques can very often be worth it.

7.5 Cognitive radio platforms: the other bits

The link between the digital hardware and the analogue circuitry is via the analogue-to-digital converter (ADC) and the digital-to-analogue converter. To achieve the performance required for a cognitive radio, not only must the DACs and ADCs have an enormous dynamic range, and be able to operate over a very wide range, extending up to many GHz, but in the case of the transmitter they must be able to handle significant levels of power. A very good and very practical paper by Bin *et al.* on ADCs discusses the range of ADCs that exist, and their relative performances [6].

The cognitive radio may also contain a whole host of additional devices, such as those used for the purposes of sensing. One such device is a global positioning system (GPS). GPS is a navigation and precise-positioning tool and is integrated into many products already. Other sensor devices which could be included are temperature, light and pressure sensors as well as accelerometers. All these devices can have small-form factors and could be integrated into a cognitive radio. Accelerometers are used in the iPhone to establish the orientation of the phone and to accordingly rotate the images on display. Any such additional devices will need to be integrated into the circuit.

7.6 Conclusions

There is much work involved in designing and building a cognitive radio. The designer needs to decide which knobs will be created (i.e. select the actions that will be allowed), determine what meters are needed within the radio, decide how to interface to meters which are remote and select the decision-making processes that determine the actions. The designer needs to decide if learning is to be incorporated and, if so, what techniques should be used. The designer needs to choose some kind of knowledge representation scheme that supports the flow of information around the radio, and to and from the outside world. The designer needs to account for security, which may include some physical-level security in the hardware as well as software solutions. The designer needs to account for higher-level protocols which may determine how networks of cognitive radios interact and communicate. The designer then needs to map this complex system to suitable digital hardware and an analogue frontend. That mapping process is likely to be challenging as the hardware may include multiple types of processor or multi-core systems and software-controlled analogue circuitry. Hence this chapter is just one anchor point in this complex design process.

References

1. T. G. Noll, Application specific eFPGAs for SoC platforms, in *IEEE International Symposium on VLSI Design, Automation and Test, VLSI-TSA, 2005*, 27–29 April 2005, p. 28.
2. L. Bélanger, Combining DSP and FPGA in next-generation multi-mode wireless handset designs, 2007. Available at: http://www.dsp-fpga.com/articles/belanger/2007/06/ (accessed 28 June 2008).
3. Q. Zhang, A.B.J. Kokkeler and G.J.M. Smit, Cognitive radio design on an MPSoC reconfigurable platform, in *Proceedings of the 2nd IEEE International Conference on Cognitive Radio Oriented Wireless Networks and Communications, CrownCom 2007*, 1–3 August 2007. Orlando, FA: IEEE Communications Society, pp. 187–91; also published in *Mobile Networks and Applications* **13**:5 (2008), 424–30.
4. J.H. Reed, *Software Radio: A Modern Approach to Radio Engineering*, Prentice Hall Communications Engineering and Emerging Technologies Series. Saddle River, NJ: Prentice Hall, 2002.
5. G. G. Ruvio, M. J. Ammann and Zhi Ning Chen, Wideband reconfigurable rolled planar monopole antenna, *IEEE Transactions on Antennas and Propagation*, **55**:6 (2007), 1760–7.
6. L. Bin, T.W. Rondeau, J.H. Reed and C.W. Bostian, Analog-to-digital converters, *Signal Processing Magazine, IEEE*, **22**:6 (2005), 69–77.

8   Cognitive radio regulation and standardisation

8.1 Introduction

To discuss regulation and standardisation in the context of cognitive radio is a challenge. Currently there are almost no regulations or standards in place for cognitive radio, as cognitive radios are still very much a thing of the future. Hence this chapter is more about classifying the general types of regulations that may be needed and the standards that are emerging than discussing what is already in place. In reality there is a wealth of regulatory issues that relate directly, indirectly or just 'kind of relate' to cognitive radio. Chapter 1 explored the role of cognitive radios in delivering new ways of managing the spectrum and looked at applications in the military, public safety and commercial domains. The new spectrum management regimes and the various potential applications may each give rise to the need for new regulations, some of which are specifically related to cognitive radios and some of which are related to creating the kind of environment in which cognitive radio applications can thrive. The purpose of this chapter, therefore, is to give a broad sense of what those issues might be, as well as to describe the current status of the standardisation efforts.

8.2 Regulatory issues and new spectrum management regimes

Much of the discussion about 'regulations for cognitive radio' is about 'regulations for new spectrum management regimes in which cognitive radios can operate'. In Chapter 1, three different new spectrum management regimes were described. We begin by reintroducing these briefly before going on to look at the types of regulatory policies that may be needed.

The first regime of interest is dynamic spectrum access. Recall that dynamic spectrum access simply means that no static assignments of frequencies are made. Instead cognitive radios dynamically access whatever free spectrum exists. As emphasised in Chapter 1 and in subsequent chapters, the use of the term 'cognitive radio' is often synonymous with dynamic spectrum access.

The second spectrum management regime is a market-driven one under which the spectrum consumers are granted exclusive usage rights to spectrum. In this case we spoke about technology and service-neutral licences. In the very broad interpretation of technology and service neutrality that we are using, a general technology and service-neutral licence will not restrict the owner of the licence to deliver any specific service or use any specific technologies and will facilitate both complete *change of use* as well as *change of ownership*. In these kinds of regime cognitive radio can play a role in dynamically sculpting the transmission profile so as to coexist with neighbours. Cognitive radios have also a potential role to play in any real-time spectrum trading which results from such regimes.

The third spectrum management approach discussed in Chapter 1 is based on an advanced commons model. In a commons regime all users are unlicensed as no users have licences which give them priority access or exclusive usage rights to spectrum. Advanced commons models move beyond using simple and universal power control methods to mitigate against interference. As emphasised in Chapter 1, a cognitive radio has the level of sophistication needed for the type of etiquettes that would be useful in a more advanced commons regime.

We can think of two broad classes of regulation that may result from the above approaches to spectrum management. The first type of regulation relates mainly to dynamic spectrum access and can be classified as being tightly coupled and dependent upon the existence of cognitive radio. The second type of regulation relates mainly to technology-neutral and service-neutral regimes as well as advanced spectrum commons regimes. These new regulations are emerging, irrespective of the existence of cognitive radio, and therefore are not solely dependent on cognitive radio. They are of interest in this book because cognitive radio is well placed to be a key technology within these regimes.

8.2.1 Spectrum management in action

To understand the new regulations that are needed it is helpful to look at current spectrum management practices. Spectrum management was introduced in Chapter 1 and the mechanics of spectrum allocation and spectrum assignment were described. These activities are part of the wider spectrum management process which includes spectrum planning, spectrum authorisation, spectrum engineering, and spectrum monitoring and compliance.[1]

1. **Spectrum planning** involves the allocation of portions of the frequency spectrum to specified uses in accordance with international agreements, technical characteristics and potential use of different parts of the spectrum, and national priorities and policies. This process is performed on both an international and national basis and, broadly speaking, international bodies tend to set out high-level guidance which national bodies adhere to in setting more detailed policy.

2. **Spectrum authorisation** involves granting access, under certain specified conditions, to the spectrum resource by various types of radio communication equipment. So far in the book we have spoken about spectrum assignment.[2] An assignment is in fact *an authorisation given for a party* to use a radio frequency or a radio frequency channel under *specified conditions*. Certification processes also fall within the general area of authorisation. Part of the authorisation work involves the certification of equipment which draws on spectrum engineering

---

1 See [1] for further discussion of regulating for radio spectrum management. The document by Cave *et al.* is part of an ICT Regulation Toolkit. In order to respond to developing countries' need for practical, relevant guidance and assistance in an ever-changing environment, in late 2004, infoDev, in cooperation with the International Telecommunication Union (ITU), began the development of an ICT Regulation Toolkit, a hands-on, web-based update and expansion of infoDev's successful Telecommunications Regulation Handbook of 2000. The Toolkit is intended to assist regulators with the design of effective and enabling regulatory frameworks to harness the latest technological and market advances. Its most prevalent themes are the impact of changing technology, the role of competition, and the regulatory implications of the transition from traditional telephony to next generation networks (NGNs).

2 This was defined in Chapter 1 – spectrum assignment happens at national level and refers to the final subdivision of the spectrum in which the spectrum is actually assigned to a specific party for use.

work as mentioned below. We should note that authorisation refers to authorising the entity that will make use of the spectrum to deliver the service of interest as well as the authorising of the equipment that will be in use.

3. **Spectrum engineering** involves the development of electromagnetic compatibility standards for equipment that emits or is susceptible to radio frequencies. The mutual interaction of radio and electrical products is known as *electromagnetic compatibility* (EMC). Spectrum engineering feeds into both the planning and the authorisation process.

4. **Spectrum monitoring and compliance** involves monitoring the use of the radio spectrum and the implementation of measures to control unauthorised use. Policing and enforcement procedures need spectrum monitoring services to identify problems and verify compliance with rules.

Dynamic spectrum access regimes, technology and service neutrality licensing and advanced commons models will have an impact on these four areas of work, each of which is now dealt with in turn.

8.2.2 Spectrum planning

The new spectrum regimes under discussion here challenge current spectrum planning practices in one very fundamental way. Whether discussing dynamic spectrum access, technology and service neutrality licences or advanced commons models, there is a move away from coupling of spectrum uses or spectrum services with specific frequency bands.[3] There are already examples of where spectrum has been auctioned with more open guidelines about how it should be used (e.g. 700 MHz auctions in the USA in early 2008). The spectrum regimes under discussion now push this approach further and in essence, in their ideal implementation, could be classified as regimes which no longer have a need for spectrum allocation (i.e. allocation of uses).

---

3 In fact technology convergence in general is also pushing this, as mobile phones now deliver TV, and telephony services are available over wireless LAN devices, etc. Hence the simplistic idea that it is possible to make statements like 'service X is delivered in frequency band A and service Y is delivered in frequency band B' is already much eroded.

A dynamic spectrum access approach is about communicating in unoccupied spectrum, with whatever waveforms suit the available space, and delivering whatever services may be appropriate. Currently, the provision of wireless broadband services tends to suit more persistent and long-term white spaces, whereas delay-tolerant networking applications are ideal for white spaces that are more bursty and short-lived. The duration and pattern of the white spaces may change over time, requiring an associated change in the services on offer. Technology and service-neutral licences for exclusive usage rights, even in the name, decouple services from being allocated to particular frequency bands. And advanced commons regimes do not need to specify usages – though the nature of the commons regime in use may lend itself to the provision of particular services (e.g. high-speed multimedia downloads over short distances) over others. This, however, does not necessitate tight stipulation of services.

The result of the decoupling of service and frequency band means that, rather than debate which services should be delivered in which frequency bands, new regimes actually centre the spectrum planning debate around which spectrum management approach should be used in which bands. Should dynamic spectrum access be allowed and if so does this mean unlicensed access should be permitted in licensed bands? Should we actually move completely to market-based approaches or should we adopt a commons model? The commons versus the exclusive usage rights debate is a very hot debate. There are those who advocate for a completely open approach to spectrum and see the commons as the perfect solution. They believe spectrum scarcity is an artefact of current spectrum management regulations rather than a real issue and hence believe a commons model would be suitable for dealing with spectrum needs. There are those who argue for complete market-based approaches and promote exclusive usage rights. The only space for a commons approach in the latter regime is some kind of private commons. The exclusive usage rights advocates also tend to shun any notion of sharing using underlays or overlays; exclusive means exclusive.

It is unlikely that any one spectrum management approach will win out. The future will consist of a patchwork of all spectrum management techniques mentioned thus far, including administrative approaches. How much spectrum is devoted to each approach is the key question. Ofcom in the UK, for example, are pushing for widespread adoption of market-based approaches. They are aiming for over 74% of spectrum to be managed in this manner. They have carried out extensive analysis of the need for larger allocations of commons spectrum but have deduced that an increase from 6% to 7% is adequate. While regulators in general have to consider incumbents and various legacy systems, as licences expire or services become outdated there are opportunities to refarm spectrum and target it towards preferred approaches.

One other topic that features in a spectrum planning discussion is the notion of the universal cognitive control channel. Should some kind of universal cognitive control channel or set of control channels be planned? The control channel concept was introduced in Section 3.5 and discussed again in the context of sensing in Chapter 4. There are clear advantages to using a control channel and, in some cases, reconfiguration will not be possible without such a service. However, whether it is necessary to explicitly plan for a universal control channel is still an open question. There are issues about what frequency bands should be used, whether multiple control channels are needed, whether it is a waste of 'good spectrum' in the first instance, whether congestion would be so great that it would be rendered useless, as well as the all-important question of who controls the control channel? The question of the need for a control channel is by no means a trivial question. It can, perhaps, only be answered as the cognitive radio field matures further.

8.2.3 Spectrum authorisation

Spectrum authorisation follows the planning phase. As already mentioned, spectrum authorisation involves granting access *under certain specified conditions* to the spectrum resource. We therefore need a process for deciding who is granted access (assigned the spectrum) and we need the specified conditions. The mechanisms for granting access range from administrative approaches involving first-come-first-served and beauty contests to market-based approaches involving auctions. What is of interest in this section, however, are the *certain specified conditions*.

Hence, the conditions for the different management regimes are the focus of the following sections.

Conditions for dynamic spectrum access

The first question that arises when attempting to specify conditions for DSA operation is whether licensed or unlicensed approaches should be used. Much of the discussion of dynamic spectrum access is based around the understanding that the cognitive radios will be unlicensed devices. This is not necessarily the only way as there are also licensed options for regulation. Both are considered here, beginning with an unlicensed approach and then looking at a licensed approach.

Unlicensed Unlicensed cognitive radio systems would permit access to the spectrum, subject to 'doing no harm' to existing users. We have come across this concept already, described in primary and secondary user terminology. The primary user is the licensed user and the secondary user in this case is the unlicensed cognitive radio which can only access primary user spectrum when the primary user is absent.

To understand the regulations that might be needed for this kind of use, it is helpful to look at an existing system which is similar. A good example is in the 5 GHz bands in which there is spectrum sharing between radar and wireless access systems. Sharing criteria specified by FCC rules (47 C.F.R. 15.407), define a *channel availability check time*, a *channel move time* and a *non-occupancy period*.

- The unlicensed device must first check if there is a radar system already operating on the channel before it can initiate a transmission on a channel. The unlicensed device may start using the channel if no radar signal with a power level greater than a given threshold is detected within 60 seconds (i.e. the channel availability check time).
- After a radar's presence is detected, all transmissions shall cease on the operating channel within 10 seconds. Transmissions during this period shall consist of normal traffic for a maximum of 200 ms after detection of the radar signal. In addition intermittent management and control signals can be sent during the remaining time to facilitate vacating the operating channel and finding a potential new channel on which to operate.

- A channel that has been flagged as containing a radar system, either by a channel availability check or in-service monitoring, is subject to a non-occupancy period of at least 30 minutes. The non-occupancy period starts at the time when the radar system is detected.

The general framework for radio behaviour, described in these bullet points, is what is of interest here. The framework is fleshed out with actual figures for the various timings and signal detection levels involved. As can be seen, the concept of *acceptable interference* or its opposite, *meaningful interference*, is built into the example. The primary user in this scenario will experience some interference. The question is how much is acceptable or, using the alternative phraseology, how to keep it at a level that is not meaningful! In the radar case the regulators have established that what is acceptable is 200 ms of normal traffic followed by intermittent management and control signals. Note also that there is no protection from interference for the unlicensed device.

The framework for radar/wireless sharing may prove useful for dynamic spectrum access. However there are specific challenges in dynamic spectrum access that are more complex and need further exploration. The detection of the primary user is an example of this. In Chapter 4 the various options for determining spectrum occupancy were discussed. The first option is to use the cognitive radio to sense available spectrum. Sensing, as we saw in Chapter 4, can be carried out at an individual node level or as part of a collaborative effort. Regulators may decide to stipulate which approaches should be used. However, this type of approach can lead to problems, and any regulations should simply stipulate the levels of signals to be sensed and not the sensing method. This is very much in line with the philosophy that emerges again and again in these spectrum management regimes – regulate for the goal rather than how to achieve the goal.

There are many who are nervous about the ability of radios, either individually or using collaborative techniques, to detect empty spectrum accurately. Hence an alternative to sensing is to use database-driven techniques in which a centralised and accessible database gives up-to-date information about spectrum occupancy and any other relevant information relating to how a cognitive radio should configure itself to make use of a spectrum hole. This approach is seen by some as very static and as a remedy to this there are also suggestions that a combination of spectrum monitoring and database techniques could be used. The spectrum monitoring output can be used to create a more 'real-time' database. This caters for quicker and short-term spectrum occupancy variations and ever more dynamic reconfiguration. As spectrum monitoring is needed for policing and compliance this option may make sense. Database and spectrum monitoring approaches also open up the opportunity for new service industries.

As a point of interest, in the USA the FCC is stipulating that databases are used in combination with sensing techniques to reliably detect the white spaces in the TV bands (see Appendix A).

Returning to the general discussion regarding a framework for dynamic spectrum access, the definition of acceptable or meaningful interference or what is meant by 'doing no harm' also needs to be considered. Again this is a much more complex issue for dynamic spectrum access than in the radar/wireless case. We will return to this later in this section as defining interference is of relevance in more than just the dynamic spectrum access case.

Licensed A licensed approach to dynamic spectrum access is also an option. One suitable mechanism for licensing is *a secondary licence without the specific agreement of the primary licensee*. This should not be confused with the term secondary user (which in most cases means secondary unlicensed user that operates around the primary licensed user). Secondary licensing is a common method for increasing spectrum utilisation. In this case, secondary licensees are granted access to the spectrum on a non-interference, non-protection basis and are generally constrained to much lower powers than the primary user. Radio microphones that use spectrum primarily used by UHF TV broadcast are examples of secondary licensed entities.

There would, however, be differences between current secondary licence approaches and ones suited to cognitive radios. In the current case the secondary licensees operate at frequencies specified by the regulators. For secondary cognitive radio licences, regulators could specify one or more spectrum bands together with licence conditions for each band. So, for example, there could be licensed underlays allowed or there could be a range of bands that are specially set aside for dynamic spectrum access which are only ever occupied by cognitive radios on a first-come-first-served basis. Hence it would be up to the cognitive users to share the bands set aside for their use making sure that in doing so interference conditions were not violated (i.e. their collective emissions within the assigned bands would not be so strong as to affect other users in the neighbouring bands). The following section on conditions for technology and service-neutral licences deals with how interference is defined and does so in a manner that is also useful for dynamic spectrum access regimes.

Conditions for technology and service neutrality

Defining spectrum usage rights for technology and service-neutral licences is challenging. Spectrum usage rights, exclusive usage rights, spectrum property rights, whatever label is used, are difficult to define because, as pointed out already, electromagnetic spectrum cannot be boxed in. In the main, the approach to defining usage rights has been to attempt to define the parameters associated with some kind of *packet/bundle/block* of spectrum with the end aim of making the definition tight enough so that:

1. the user of the *packet/bundle/block* has a clear set of entitlements
2. neighbouring *packets/bundles/blocks* have a clear set of entitlements
3. it becomes possible to trade a *packet/bundle/block* or multiples thereof in some kind of market system.

The temptation is to use frequency, space and time to define a block of rights, i.e. a spectrum consumer is given a frequency range associated with a geographical area or space for a given time period and it can deliver the services it so chooses to its users. From exploring interference issues in Chapter 3, we know that electromagnetic spectrum spreads beyond its frequency and spatial borders. We reuse the summary interference image from Chapter 3 here again to remind ourselves of the types of interference that must be taken into account by the regulator. Figure 8.1

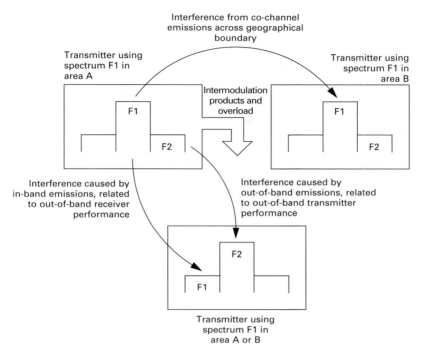

Fig. 8.1. Different ways in which a signal can spill into neighbouring frequencies and other geographical areas.

helps underline the complexities involved in making regulations relating to interference by showing the spillage that occurs from one frequency to another and from one geographical area to another. Hence the block of spectrum usage rights cannot just be defined in terms of frequency range, geographical space and time; instead something that captures how much spillage is allowed must be used. So rather than have a *frequency/space/ time-period* block we might instead have a *within-frequency-band-signal-strength-limit/beyond-frequency-band-signal-strength-limit/with in-space-signal-strength-limit/beyond-space-signal-strength-limit/time-period* block of rights.

The suggested way of dealing with all of this is to use some kind of *emission mask*. An emission mask specifies an emission profile within which a transmitter must stay. An example of a simple mask is an *equivalent isotropic radiated power* (EIRP) mask. Haslett [2] defines this well. When a signal is transmitted via an antenna, that signal has a certain transmit power, P_t, and the antenna will typically have a gain that amplifies the signal, G_t. It is common to regard $P_t + G_t$ as a single entity as it is the sum of the two terms that is significant rather than the individual terms of transmitted power and transmitting antenna gain. The entity is the EIRP. The same power would be received if a real power equal to the EIRP was transmitted through an isotropic antenna as would be received through the actual configuration. Thus the same power would be received if a power of 50 dBm was transmitted through an antenna of 15 dBi gain as if a power of 40 dBm was transmitted through an antenna of 25 dBi gain. The EIRP in both cases is 65 dBm.

To look at some alternatives to EIRP we can call on the Ofcom 2006 consultation document [3]. The main metric used in the Ofcom masks is *power flux density* (PFD). This is another kind of power measurement. The formal definition of power flux density is the power crossing unit area normal to the direction of wave propagation. It is possible to talk about the power flux density in a location i.e. as experienced by a receiver in a given location due to a transmitter or to multiple transmitters. So, unlike an EIRP mask, which more or less specifies the settings of a given transmitter (i.e. power level should be set to X to get EIRP of Y), the power flux density can be thought of as focusing on the experience of the receiver. And more importantly the power flux density at a point is based on the emissions from all transmitters that may exist in a given area. A good example of the use of power flux density is in the suggested Ofcom mask for dealing with the spillover of power between geographical areas.

- The aggregate power flux density (PFD) at or beyond [definition of boundary] should not exceed X dBW/m^2/[reference bandwidth] at any height up to H m above local terrain for more than $P\%$ of the time.

The term aggregate power flux density emphasises that the mask focuses on total power flux density of all of the transmitters in a geographical area A that operate on frequency F1. Hence transmitters on the same frequency, F1, but in geographical area B will not expect to experience a signal that is greater than a given flux density and, based on that information, can situate themselves out of harm's way of co-channel interference.

If the spectrum consumer decides to change the number and pattern of distribution of the transmitters it is using, it has to ensure the new configuration produces the same PFD levels at the boundaries. Masks can also be defined to deal with frequency boundaries. The following mask is an example of one which stipulates how much spillover into out-of-band frequencies is allowed by the transmitter.

- The out-of-band PFD at any point up to a height H m above ground level should not exceed X dBW/m^2/MHz for more than $Y\%$ of the time at more than $Z\%$ of locations in any area A km^2.

What this means is that any receiver in the area who is operating at frequencies that fall within the out-of-band frequency range of the transmitters in the area will not be expected to deal with unwanted signals of greater than a particular strength. The final mask sets up a level of expectation of the kind of adjacent channel interference that would be expected by a given receiver.

- The in-band PFD at any point up to a height H m above ground level should not exceed X dBW/m^2/MHz for more than $Y\%$ of the time at more than $Z\%$ of locations in any area A km^2.

The Ofcom report [3] recommends the use of all three masks. The crucial values of the various parameters are missing in the definitions used here but the Ofcom report does suggest some values, and subsequent work has been carried out in further investigating these usage rights. The masks here are merely examples to give a flavour of the type of masks which may prove useful.

Though the masks have been discussed in the context of technology and service-neutral licences, similar definitions for interference limits could be used in dynamic spectrum access contexts. It is possible to determine suitable power flux density levels that would define acceptable interference levels for dynamic spectrum access operation. This may necessitate the incorporation of some of the timing dynamics into the masks. So, for example, rather that talking about '$Y\%$ of the time' one might instead refer to 'for up to Y time after the return of the primary user' in the case of the unlicensed secondary users.

One point of importance that should be stressed before moving on from the discussion on interference is the idea of 'a starting point only'. Very many of the discussions around usage rights support regulations that leave open the option for negotiation between neighbours. Hence the interference rules are seen as a starting point or a basis on which to work. Whether you think of it as Coasean bargaining[4] or some other kind of negotiated use, regulating in a manner that facilitates this is essential.

Finally, when talking about conditions for technology and service-neutral licences, some mention should be made of trading-related issues. As we learned already, the exclusive usage rights /technology and service-neutral approaches lend themselves well to spectrum trading. This was discussed briefly in Chapter 1. Of particular interest is the role that cognitive radio can play in spectrum trading in secondary markets especially in more real-time or 'just-in-time' spectrum trading. While this may be a long-term vision, it is worth noting that there are many regulations already in place for spectrum trading, albeit on much larger scales and at much slower paces. In Chapter 6 of *Essentials of Modern Spectrum Management* [4] a selection of the spectrum trading rules in place in different countries around the world are described. While none of the rules cater for real-time or just-in-time trading they do provide useful background information and a starting point. Real-time trading would require an evaluation of various auction mechanisms to determine whether they could be used for very-short-term trades. The implementation of an automated trading process depends upon finding mechanisms that ensure transaction costs far less than the value of the trades.

Conditions for advanced spectrum commons

The mechanism for regulating commons regimes has to date involved specifying a low-level EIRP mask which limits range and hence the possibility of interference to other users. While useful, this does limit the breadth of applications that can be used in these bands. There is no

---

4 Nobel Laureate Ronald Coase was a leader in criticising spectrum regulation and calling for tradable property rights. The 'Coase Theorem' basically states that well-defined property rights and low transaction costs allow parties to bargain to reach efficient outcomes.

ability to react dynamically – if I am in the middle of a desert with no one around other than the node with which I am communicating, why shouldn't I use higher power? Other rules are therefore needed.

Just as in the case of dynamic spectrum access and technology and service-neutral licensing there are many ways in which a more advanced commons could be delivered. As a case study we will look at the Ofcom Licence-Exemption Framework Review. This consultation on the framework for managing spectrum used by licence-exempt devices was published in 2007. It captures a number of the issues of relevance and gives an indication of how a more advanced spectrum commons might be regulated. In the report there are two main suggestions for operating a commons which are relevant here. The first is the suggested use of what Ofcom call *politeness protocols*. We used the term 'etiquette' in Chapter 1 and it essentially means the same thing. The idea is that the high-level politeness rules would be defined by the regulator and the regulator would also need to authorise the use of any standardised polite protocols within a spectrum commons. There are many different kind of politeness protocols that could be used. A simple 'listen-before-talk' is listed in the report. But, as discussed in Chapter 1, protocols which involve bargaining and 'dynamic redistribution of the wealth' could also be used. At the time of writing the details of the politeness rules or the politeness protocols have not been specified.

The second concept of interest in the Ofcom report is the concept of multiple types of commons – i.e. not all commons operate under the same rules. The idea here is to define different classes of spectrum commons for communications ranges of the order of up to metres, tens of metres, and hundreds of metres and in which different politeness rules apply. While the example from the Ofcom report on the license-exempt framework is, as stressed already, just one way of doing things, the example does serve to illustrate the types of regulations which may emerge and which may have a bearing on cognitive radio.

Dynamic authorisations
The type of conditions detailed thus far could form the basis on which flexible authorisation is granted to an entity. The cognitive radio itself can

enable further more dynamic approaches to authorisation. Whatever the exact configuration of a cognitive radio, it is envisaged that some sort of policy engine will provide a means of checking for the 'all clear' before the radio configures itself in order to engage in a communication session. With this kind of approach it is possible to consider authorisations that are linked to a very dynamic policy database that can be changed centrally and can come into effect seamlessly when the radio accesses the database for guidance/instructions.

Time-limited leases, suggested by Chapin and Lehr [5], are another more dynamic means of authorising access to spectrum. Leases are extended by delivering a certificate to a device. According to Chapin and Lehr, time-limited leases behave just like the time-out programmed into trial versions of software packages. In this case, the time limit is built into a radio device. If the time limit is reached and no extension message is received, the radio reduces its behaviour as required or potentially halts transmission entirely. The leases can be extended by the regulator depending on circumstances. The lease can be issued using some kind of digital certificate that is encrypted and/or cryptographically signed to assure that only the responsible authority is able to extend the lease. Certificates can be sent to the device proactively, or retrieved automatically by the device as the end of the current lease approaches. Time-limited leases may be useful for a whole myriad of reasons, one of which will become clear in the next section.

8.2.4 Spectrum engineering

We also noted that certification processes fall under the heading of spectrum authorisation, as those who are granted access to the resources and their equipment must be certified. Spectrum engineering involves the development of electromagnetic compatibility standards for equipment that emits or is susceptible to radio frequencies. Device certification is the process where a radio must be shown to comply with standards (interference and safety regulations) before sale. Different countries can have their own standards and the ITU also provides equipment standard regulations and recommendations to its members. Certification, in fact, is a step that must be carried out in order for authorisation to be granted.

Currently, the manufacturer tests a device in all of its operating modes, and measures the emissions to show that it never violates the regulations under which it must operate. Provided the tests are successful, the regulator grants the manufacturer the right to manufacture, sell and operate the device as certified. This approach suits single-purpose hardware used to support narrow ranges of wireless applications. Cognitive radios will bring many challenges to the certification process. As we have seen throughout the book, the fact that a cognitive radio is designed to adapt to the operating environment in which it finds itself typically means that the cognitive radio has a much wider range of modes of operation. Added to this, cognitive radios can learn and can be said to evolve. Networks of cognitive radios can exhibit emergent properties. There is thus a requirement to understand how the radio behaves 'in the wild'. Hence the certification of these devices will require very different tests than are currently in operation.

There are some indications of how certification may be tackled. Very many cognitive radio implementations are software radio based as discussed in Chapter 7. There are regulations in place in certain jurisdictions for dealing with the certification of software radios and, as software radios have many of the same characteristics as cognitive radios, they can provide some insight into how certification of cognitive radios can be handled. For reference purposes the FCC designates radios as software radios if 'the software is designed or expected to be modified by a party other than the manufacturer and would affect the listed operating parameters or circumstances under which the radio transmits'. According to FCC regulations the original version of the software radio has to be certified with the equipment on which it will run. The FCC does not certify modifications made by third-party software vendors. All modifications are the responsibility of the manufacturer whose hardware/software radio bundle was originally certified. Thus, the current rules require independent radio software providers to work with the equipment manufacturers.

There is ongoing research into the certification processes for cognitive radio. There are suggestions for adopting a very modular approach to certification so that different modules of the cognitive radio can be certified in different ways and to different levels of accuracy. The policy engine can also be exploited to perhaps make more light work of certification.

There may be a way of ensuring that no policy non-compliant behaviour is possible resulting in reduced need for a detailed analysis of all modes of operation of the radio itself. The time-limited lease approach mentioned above could perhaps also mitigate against risks. As Chapin and Lehr [5] suggest, 'A regulator, faced with a device too complex to test thoroughly, can certify it for sale and operation knowing that it is easy to recall if it misbehaves in the field. As long as the device behaves safely, the lease will be freely extended for additional time periods.' In this case the cognitive radio is itself a solution to the problem of certification of cognitive radio. However, like most issues in this chapter, more work is needed.

8.2.5 Spectrum monitoring and compliance

Even if devices are properly certified it is still necessary to perform some kind of monitoring to check that operators and spectrum consumers are behaving within the limits of their authorisations. Monitoring is also important in resolving any cases of intra-national or international interference that might arise. Hence when thinking about regulations for spectrum management, regulations for monitoring and compliance must also be considered. Monitoring is standard practice for validating information on legitimate users, evaluating real levels of usage of the spectrum as well as for investigating compliance.

The question for this chapter pertains to what is new in monitoring for cognitive radio? The answer is threefold. Firstly, many of the new regimes may require more monitoring, both as a safety net for those anxious about new regimes as well as a means of ensuring compliance. Secondly, the cognitive radio may enable the use of new monitoring techniques. For example, there may be an equivalent digital action to 'stopping a driver and asking to see a driver's licence'. Thirdly, many of the functions of the monitoring process, such as spectrum occupancy measurements, may in themselves become services, not for the regulator but for the wider user base. Distributed monitoring services, as mentioned already, may provide crucial information in dynamic spectrum access regimes or in any kind of automatic configuration or organisation of a radio or network.

8.3 Cognitive radio applications and regulations

The regulatory issues discussed thus far have been discussed in terms of general spectrum management practice, and quite wide-ranging issues have arisen. Regulations can also be discussed in terms of cognitive radio application areas and these bring further issues to the fore. Some of the applications touched on in Chapter 1 do not require new regulations. A cognitive radio or a cognitive network could be used in any current system to improve system performance: for example, a cognitive WiMAX system could have features that allowed different WiMAX basestations to auto configure. Applications of cognitive radio in the cellular world could allow cellular users to make more dynamic and efficient use of the spectrum for which they already have licences. The military have freedom to cater for new applications within their own system. Some public safety applications, however, do lead to the need for new regulations especially in cases in which public safety and commercial applications may coexist.

The so-called 'D Block' spectrum that was auctioned unsuccessfully in the USA in early 2008 is a good case study for this discussion. Recall from the discussion on the digital dividend in Chapter 1. In the USA TV stations operate on 6 MHz channels designated 2 through 69, namely 54–806 MHz in the VHF and UHF portions of the radio spectrum. To replace the current channels with digital TV channels requires less bandwidth and all the digital TV stations contained within channels 2–51. The portion of the spectrum channels 52 through 69, namely 698–806 MHz, has been reallocated to other services. The FCC auctioned these frequencies in early 2008. The auctions were colloquially termed the '700 MHz auctions'. The entire spectrum range was divided into different blocks, and different conditions and reserve prices were placed on each.

The D Block was the spectrum that was set aside for public safety. The idea behind the D Block was the establishment of a public/private partnership to promote rapid construction and deployment of a nationwide, inter-operable broadband public safety network that could also be used commercially. The D Block did not reach its reserve price and was withdrawn. It is not clear why the D Block auction failed. There is a wide range of possible reasons. Issues with the public/private partnership concept, the reserve price, how the auction was constructed, the technical specifications, the costs of realising such a network and many more concerns could have played a role. The FCC are at the time of writing exploring how to proceed. The Second Further Notice on Proposed Rule Making – On Implementing a Nationwide, Broadband, Interoperable Public Safety Network in the 700 MHz Band (WT Docket No. 06-150 and PS Docket No. 06-229) discusses the many issues. It serves as an excellent example of the very many regulatory and policy issues that arise from sharing between commercial and public safety users.

While cognitive radio was not mentioned in the proposals for the D Block, cognitive radio does seem the perfect technology for a dual-use public safety/commercial network. It has the potential for finding solutions to automatically prioritise public safety communications over commercial uses on a real-time basis and assigning highest priority to communications involving safety of life, and property and homeland security. However, the fact that cognitive radio has great potential for delivering this kind of public safety/commercial mix but is, as yet, an immature technology, may have played a role in the failure of the D Block ideas to take off.

Other public safety applications just require more specific versions of regulations already discussed. A good example is the proposal by Jesuale and Eydt [6]. They suggest pooling public safety, business/industrial and federal/non-military spectrum with the use of dynamic spectrum access. These bands are not contiguous but the treatment of them as a pool through the use of cognitive radio techniques opens greater gains and efficiencies. Spectrum planning would have to take this into account, authorisations of users would have to allow for the dynamic spectrum access approach across the bands and, given that public safety is the issue, more stringent definitions of acceptable interference would no doubt arise.

8.4 Standards and international activity

Cognitive radio is the focus of a number of standardisation activities. Some of the activities are in a very early state while others are more advanced. To date some key international bodies have only gone as far as putting cognitive radio on the agenda while other groups have been discussing cognitive radio in detail for a number of years. The following sections of the chapter form a summary of the main activities in the field.

8.4.1 The work of the SCC 41

One of the most well-known activities ongoing relating to cognitive radios and networks is the work of the IEEE Standards Coordinating Committee 41 (SCC 41) Dynamic Spectrum Access Networks (DyS-PAN). SCC 41 sponsors standards projects in the area of dynamic spectrum access networks and provides coordination and information exchange between and among standards-developing activities of the IEEE.

SCC 41 was born out of what is known as the IEEE P1900 Standards Committee. The latter was established in 2005 jointly by the IEEE Communications Society (ComSoc) and the IEEE Electromagnetic Compatibility (EMC) Society. The objective of this effort is to develop supporting standards dealing with new technologies and techniques being developed for next-generation radio and advanced spectrum management. On 22 March 2007 the IEEE Standards Board approved the reorganisation of the IEEE 1900 effort as Standards Coordinating Committee 41 (SCC 41), Dynamic Spectrum Access Networks (DySPAN). The IEEE Communications Society and EMC Society are the sponsoring societies for this effort. SCC 41 has a number of working groups, namely 1900.1, 1900.2, 1900.3 and 1900.4, all tasked with different projects. The following sections outline the activities of each of these groups.

IEEE working group 1900.1

The 1900.1 working group is mainly a definitions group. It is probably clear from the outset of this book that there are imprecise and multiple definitions for cognitive radio. The P1900.1 group is attempting to provide technically precise definitions and explanations of key concepts in the fields of spectrum management, cognitive radio, policy defined radio, adaptive radio, software defined radio, and related technologies. The document being produced by the group is extensive and goes beyond simple, short definitions by providing amplifying text that explains the various technologies. The document also describes how these technologies interrelate and create new capabilities while at the same time providing mechanisms supportive of new spectrum management paradigms such as dynamic spectrum access.

IEEE working group 1900.2

The 1900.2 looks at recommended practice for the analysis of in-band and adjacent band interference and coexistence between radio systems. This recommended practice provides technical guidelines for analysing the potential for coexistence or in contrast interference between radio systems operating in the same frequency band or between frequency bands. This is the kind of analysis that has the potential to feed into the definition of regulations for interference, including such discussions of spectrum usage rights as in Section 8.2.3.

IEEE working group 1900.3

The aim of the 1900.3 working group is to specify techniques for testing and analysis to be used during regulatory compliance and stakeholder evaluation of radio systems with dynamic spectrum access (DSA) capability. This relates very much to the certification issues that were discussed in Section 8.2.4. The standard being developed focuses on RF system test and certification.

IEEE working group 1900.4

The purpose of this 1900.4 group is to improve overall capacity and quality of service of wireless systems in a multiple Radio Access Technologies environment by defining an appropriate system architecture and protocols which will facilitate the optimisation of radio resource usage, in particular by exploiting information exchanged between network and mobile terminals, whether or not they support multiple simultaneous links and dynamic spectrum access. The work of this group is more about creating an environment that supports reconfiguration and optimisation of devices that best suit their needs, should they so choose.

8.4.2 IEEE 802.22

The IEEE 802.22 standard has been mentioned already in the book. However, it is worth recalling here briefly as it is one of the few standards that is under development for specific use of cognitive radio. In Chapter 1 the move to digital TV was discussed and proposals to make the TV white spaces available to unlicensed users was introduced. As a result, the IEEE 802.22 working group was set up to develop a standard for wireless regional area networks that would make use, on a non-interfering basis, of the TV white spaces. The 802.22 working group looks at physical layer and medium access control layer issues as well as spectrum sensing, geolocation/database and security issues. Aspects of the standard have been discussed where relevant during the book, in particular those relating to sensing and security.

8.4.3 International Telecommunications Union activities

The International Telecommunication Union (ITU) is responsible for the governance of spectrum on a global basis. It is not a global authority as the international rules are written by those governed by them, i.e. the member states of the ITU. These rules are administered by the ITU's Radiocommunication Bureau (BR), and conformity with the rules is based on goodwill and supported by regulations at the national level.

The Radiocommunication sector of the ITU is called ITU-R. This sector carries out various studies and adopts recommendations on radio communication matters. The ITU-R has initiated standardisation activities in the area of cognitive radio and cognitive networks. The ITU-R arranges the World Radiocommunication Conference (WRC) every two to four years to review and revise the radio regulations governing the use of the radio spectrum and satellite orbits. These radio regulations determine frequency allocations to different services as well as specifying rules for using the various bands. In the World Radio Conference held in 2007, the ITU-R studies concluded that mobile communication systems require more spectrum than was previously allocated to them and as a result new techniques for more efficient spectrum use are needed. With this in mind the World Radio Conference for 2011 (WRC-11) has an agenda item on software-defined radio and cognitive radio that will focus on regulatory measures to enable the introduction of software-defined and cognitive radio systems. While this is very welcome, this is a good example of the pace at which ITU processes advance.

Before going into more details of the ITU-R activity, it is worth dwelling for a moment on the coupling of software-defined radio and cognitive radio. These two technologies are, more often than not, coupled. This is understandable because, while cognitive radios can be constructed without needing a software radio, the main implementation techniques all revolve around programmable hardware and/or software. And a cognitive radio is often seen as an advanced version of a software radio, and software radio is seen as 'the' enabling technology for cognitive radio.

The ITU uses study groups (SGs) to focus on different technical questions. There are two study groups which are of relevance here. Measurement of spectrum occupancy is studied in ITU-R SG 1. In Chapter 1, spectrum occupancy measurements were presented in Figure 1.3 showing the unused spectrum over the particular band of interest. These types of measurements are taken very seriously and have motivated much of the work in the field of dynamic spectrum access. There are, however, many questions concerning the measurements that warrant further investigation. It is often argued that low-power spectrum occupancy is not captured by these measurements. SG 1, with this in mind, is working on the following four issues:

1. What techniques could be used to perform frequency channel occupancy measurements, including processing and presentation methods?
2. What techniques could be used to perform frequency band occupancy measurements, including processing and presentation methods?
3. How can occupancy be defined for both frequency channel as well as for frequency band measurements, also taking into account the size of the used filter and the values measured in adjacent channels?
4. How can threshold levels be defined and applied in practical situations, including dynamic threshold levels?

The answers to some of these questions impact on many of the regulations discussed in the first half of this chapter. The responses to these questions will become part of ITU-R recommendations and will be completed by 2009.

The second study group of interest is SG 5. This group is focusing on cognitive radio in mobile services. The following are the questions this group is tackling:

1. What is the ITU definition of cognitive radio systems?
2. What are the closely related radio technologies (e.g. smart radio, reconfigurable radio, policy-defined adaptive radio and their associated control mechanisms) and their functionalities that may be a part of cognitive radio systems?
3. What key technical characteristics, requirements, performance and benefits are associated with the implementation of cognitive radio systems?
4. What are the potential applications of cognitive radio systems and their impact on spectrum management?

The responses to these questions will also become part of ITU-R recommendations and/or reports, and these will be completed by 2010. As is obvious from these questions, there is overlap between various groups and their activities. The SCC 41, for example, has a well-matured focus on cognitive radio definitions.

8.4.4 ETSI activities on cognitive radio

ETSI is the European Telecommunications Standards Institute. ETSI is recognised as an official European Standards Organisation by the European Commission. ETSI produces globally applicable standards for fixed, mobile, radio, converged, broadcast and Internet technologies. ETSI has put together a technical committee to examine standardisation and development of software-defined radio and cognitive radio. This committee had its first meeting in March 2008. The committee aims to use results from EU-funded research in the area of software and cognitive radio.

8.4.5 SDR Forum

The Software Defined Radio Forum (SDRF) is an international non-profit organisation dedicated to promoting the development, deployment and use of software-defined radio technologies for advanced wireless systems. It has over 100 organisational members, mostly from industry and research fields, is well known in the industry and holds regular meetings. As software radio is an enabling technology for cognitive radio, the SDR Forum takes a strong interest in cognitive radio.

Of the various work that is ongoing in the SDR Forum, there are three groups that are relevant to cognitive radio. The first is the Regulatory Committee. This committee aims to promote the development of a global regulatory framework supporting software download and reconfiguration mechanisms and technologies for SDR-enabled equipment and services. Software download has not been explicitly discussed here but it relates to the downloading of new software for reconfiguring a radio (in line, for example, with the regulations of the jurisdiction in which the radio finds itself as well as a means of upgrading its functionality).

The second group of relevance for cognitive radio is the Cognitive Radio Working Group (CRWG), which is forming a general consensus of what cognitive radio and cognitive radio related terms mean and imply. This is important since an agreed set of definitions is fundamental to regulation of CR technology. This group has some interaction with the 1900.1 working group as they have concerns in common.

The third and final group is the Cognitive Applications Special Interest Group (CA-SIG), which is identifying wireless communication and control scenarios where cognitive radio and cognitive radio applications could provide a better solution than traditional approaches. A purpose of this group is to help identify which cognitive radio applications are likely to emerge first. While Chapter 1 details a large number of cognitive radio applications there is still much debate as to which applications will take the lead.

8.5 Conclusion

Regulations for dynamic spectrum access have a very direct bearing on cognitive radio. Regulations for technology and service-neutral regimes and advanced commons regimes are important also. While not wholly dependent on cognitive radio, a cognitive radio mindset has a lot to bring to the table when discussing these topics. Various cognitive radio application domains bring with them a variety of other regulatory concerns, most visibly in the area of public safety policy.

As cognitive radio is still in its infancy, we can only suggest broad outlines for policies. There is a tension between creating policies that are flexible enough so that the real advantages of cognitive radio can be exploited and creating policies that are restrictive enough to ensure protection for existing systems. However, there are many advances in technology and in regulatory policy thinking to be able to deal with the tension in due course. (See Appendix A for some rules that have been agreed in the USA.)

The ability to be able to freely test and trial some of the regulatory policies with real cognitive radios may be the only way to answer many of the regulatory questions that arise. Spectrum has been set aside or opportunities defined in a number of countries which support this kind of activity. The Irish Commission for Communications Regulation (Com-Reg) and the American Federal Communications Commission (FCC) are just two examples of countries which have taken this path. These kind of initiatives should be exploited to the full.

As discussions about regulatory policy proceed, there is continued international activity in exploring cognitive radio, its impact and potential standards for its use. Currently a certain amount of work is taking place in parallel but the main point is that cognitive radio is very much on the agenda.

References

1. M. Cave, A. Foster and R. W. Jones, Radio spectrum management: overview and trends. ICT 2006. Downloadable from http://www.ictregulationtoolkit.org/en/publication.1889.html
2. C. Haslett, *Essentials of Radio Wave Propagation*, Cambridge Wireless Essentials Series, Cambridge University Press, 2007.
3. Ofcom, Spectrum usage rights: Technology and usage neutral access to the radio spectrum. Office of Communication, 2006. Available at http://www.ofcom.org.uk/consult/condocs/sur/ (accessed October 2008).

4. M. Cave, C. Doyle and W. Webb, *Essentials of Modern Spectrum Management*, Cambridge Wireless Essentials Series, Cambridge University Press, 2007.
5. J. Chapin and W. Lehr, Time-limited leases in radio systems, *IEEE Communications Magazine*, **45**:6 (2007), 76–82.
6. N. Jesuale and B. C. Eydt, A policy proposal to enable cognitive radio for public safety and industry in the land mobile radio bands, in *2nd IEEE International Symposium on New Frontiers in Dynamic Spectrum Access Networks, 2007*. Dublin, 17–20 April 2007, pp. 66–77.

9   Conclusions

9.1 Introduction

This book has considered the essential issues associated with cognitive radio. Cognitive radio is a truly interdisciplinary topic. It crosses the fields of information theory, propagation studies, RF design, telecommunications, wireless networking, signal processing, artificial intelligence, cognitive science, software engineering, regulatory policies, security, application design, plus many many more. On the one hand the need for such a breadth of knowledge seems daunting and on the other hand it seems very exciting and opens up the way for new possibilities. In this last chapter an attempt is made to summarise the main points of the book.

9.2 A brief summing up

The first main point of the book is that a cognitive radio is not just a radio for dynamic spectrum access. The second is that dynamic spectrum access is a great idea and in looking at dynamic spectrum access many new paradigms for dynamic behaviour of radios come to light. The third point is that there are many potential applications on the horizon for cognitive radio as is hopefully clear from the simple mindmap in Figure 9.1.

Throughout the book the **observe**, **decide** and **act** cycle was placed at the core of the cognitive radio (Figure 9.2 recaptures the cycle). And **secure**, **build** and **regulate** were seen as the key additional activities of focus. Each of these topics warrants a few words.

9.2.1 Observe

A cognitive radio has four inputs, an understanding of the environment in which it operates, an understanding of the communication requirements of the user(s), an understanding of the regulatory policies which apply to

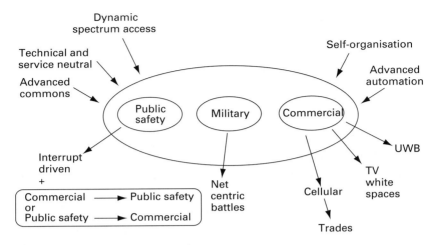

Fig. 9.1. A mindmap of Chapter 1.

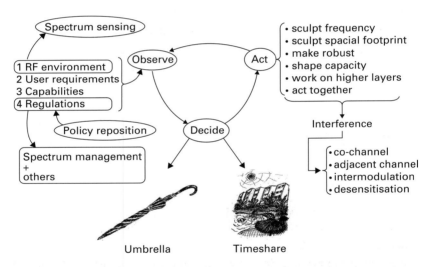

Fig. 9.2. A last look at the 'observe, decide, act' cycle.

it and an understanding of its own capabilities. Getting these four inputs is what we mean by the phrase 'observing the outside world'. The radio can get observations itself or it can turn to other radios or external entities for help. Of all the inputs the understanding of the RF environment is the most challenging and much of the effort in the field is in spectrum sensing. Nonetheless the large number of other measurements that a radio can make should not be forgotten. There is potential for creating a very rich awareness of the world.

9.2.2 Decide

Decision-making is at the heart of cognitive radio. To do this we need to represent knowledge in a suitable fashion and have mechanisms for reasoning, optimising and learning. There are multiple approaches. To date many solutions tend to focus on one method and one decision-making scenario. In reality a richer approach which encompasses a number of techniques is needed, bearing in mind that 'cognitive engines should be as simple as possible, but no simpler'. In the end all decisions are either umbrella-like or apartment time-share-like decisions. It is just a matter of having the techniques to hand for each one and feeding in the observations and constraints.

9.2.3 Act

'Taking action' in cognitive radio terms is all about reconfiguring its parameters of operation, to adjust its operating characteristics to suit prevailing conditions. It is important to think of the radio as a node in a network and to think from application layer to physical layer. From this vantage there are an enormous number of actions that can be taken. In this book actions were classified under seven broad headings, namely actions that shape the frequency content of the signal, actions that shape the spatial footprint, actions that make the signal robust for its journey, actions that increase capacity, actions that distribute resources, actions that apply to higher layers and actions that can be taken collectively. The key to understanding these actions is understanding their consequences. Actions have consequences and the main consequence is interference. The ability of a transmitter to communicate reconfiguration details to a receiver or group of receivers brings up the challenging question of a control channel. There are different ideas on how to deal with this but work still needs to be done.

9.2.4 Secure

Cognitive radios have unique security needs. There are weaknesses because of the high dependency on observations, because of the need to collaborate, because of the high levels of software involved, because of single points of failure and because of the kind of applications that are in the cognitive radio domain. The potential strength that can be brought to bear is that a cognitive radio is sophisticated enough to be able to deal with more imaginative security solutions. But even if less-imaginative ones are used, security should be included from the beginning.

9.2.5 Build

Building cognitive radios is challenging. However, it basically reduces to deciding which knobs and meters the radios should have and what 'intelligence' should be applied to feed the meter readings and other wider observations into the decision-making process in order to arrive at a setting for the knob or knobs. The tasks involved in this process can be hugely wide ranging and all have differing demands. This makes it very likely that a highly heterogenous platform will be used as the fabric of the cognitive radio.

9.2.6 Regulate

There are a great many regulations that relate to cognitive radio because cognitive radio has so many potential application areas. The main area of focus is dynamic spectrum access, but other spectrum management regimes such as technology and service-neutral regimes and advanced commons regimes matter too. The application areas may give rise to further new policies and regulations, but none more so than public safety.

All in all there is huge advancement in regulatory thinking that very much suits cognitive radio.

9.3 The future

Cognitive radio most definitely has a future. There is enormous scope for adopting at least some of the principles in this book and more. This is not without challenge. There are still open technical problems. The list is long but paramount is the technical challenge of bringing all the functionality needed for cognitive radio together in a comprehensive manner. The resulting system not only needs to be robust and reliable, but needs to be testable and certifiable. There are regulatory issues that are only beginning to be discussed. There are political issues, especially when it comes to taking a more open approach to access to spectrum and its use. There is work to be done in further fleshing out the application areas and the associated business models.

How 'cognitive' the next-generation radios will be remains to be seen. During a discussion in our research group about aims for our work, one of the team suggested we should aim for *a radio that is so cognitive, it gets bored.*[1] Whatever the level of academic folly in this, cognitive radio is never boring.

1 This comment is attributed to Keith Nolan.

Appendix A Developments in the TV white spaces in the USA

During the production phase of this book, the FCC released two reports that are of relevance to this book. At that stage it was too late to include details of the reports in the main body of the text. This short appendix addresses the issues briefly.

On 15 October 2008 the FCC released their report (FCC/OET 08-TR-1005) on the Evaluation of the Performance of Prototype TV- Band White Space Devices Phase II. The opening paragraph of the report summarises what the report shows:

The Federal Communications Commission's Laboratory Division has completed a second phase of its measurement studies of the spectrum sensing and transmitting capabilities of prototype TV white space devices. These devices have been developed to demonstrate capabilities that might be used in unlicensed low power radio transmitting devices that would operate on frequencies in the broadcast television bands that are unused in each local area. At this juncture, we believe that the burden of 'proof of concept' has been met. We are satisfied that spectrum sensing in combination with geo-location and database access techniques can be used to authorize equipment today under appropriate technical standards and that issues regarding future development and approval of any additional devices, including devices relying on sensing alone, can be addressed.

The report goes on to state that

All of the devices were able to reliably detect the presence a clean DTV signal on a single channel at low levels in the range of -116 dBm to -126 dBm; the detection ability of each device varied little relative to the channel on which the clean signal was applied. These measurements did not take into account the antenna that would be used with personal/portable devices.

In Chapter 1 of this book, the use of cognitive radio in the TV white spaces was flagged as a potential first commercial application area. The

FCC report, in confirming that it is possible to reliably sense TV signals, even though some more work is needed, is paving the way for the use of cognitive radios in the TV white spaces. As we learned in this book, spectrum sensing is a key aspect of the observe stage of the 'observe, decide, act' cycle. The report provides interesting reading for those wishing to look at the practicalities of spectrum sensing in more detail.

On 4 November 2008, the FCC voted 5–0 to approve the unlicensed use of white space (there are five FCC commissioners). This was followed by the release of what is known as a Second Report and Order, ET Docket No. 04-186 and ET Docket No. 02-380, which detail the rules under which unlicensed devices could use the TV white spaces. The rules provide for operation of both fixed and portable unlicensed devices in the TV bands. Note that the term portable rather than mobile is used. At TV band frequencies the kind of antennas needed mean you can move a device around if you need to rather than that it is actually mobile! There are a number of specific rules which are of key interest in the context of this book.

The first set of rules of interest relate to how sensing is carried out. The FCC stipulates that all devices, except personal/portable devices operating in client mode, must include a geolocation capability, provisions to access over the Internet a database of protected radio services and the locations and channels that may be used by the unlicensed devices at each location. The unlicensed devices must first access the database to obtain a list of the permitted channels before operating. The checking of the database must be supplemented with additional sensing. The report states that fixed and personal/portable devices must also have a capability to sense TV broadcasting and wireless microphone signals as a further means to minimise potential interference. However, for TV broadcasting the database will be the controlling mechanism. The report does allow for devices that will not use the database but stipulates that these devices will be much more rigorously tested and hence it appears the bar for getting such devices certified will be very high.

The second set of rules which are of interest relate to the power limitations of the devices. Fixed devices may operate at up to 4 watts EIRP (effective isotropic radiated power). Personal portable devices may operate at up to 100 milliwatts of power, but operation on adjacent channels will be limited to 40 milliwatts.

The report gives much detail about the exact TV bands in which these new unlicensed devices can operate, and the restrictions relating to co-channel and adjacent channel interference. (These topics are discussed in Chapter 3.) And it also gives details of how wireless devices are protected, among other topics. It is a very useful report as it provides an opportunity to see how many of the issues relating to cognitive radio, as discussed in this book, are translated into practical rules and guidelines.

Index

700 MHz auctions, 213
   D Block, 213

ADC, 193
antennas, 190
apartment time-share example, 38
ASIC, 171
ASIP, 172

beamforming, 19, 73, 191

cellular communications, 29
channel impulse response, 120
commercial, 25
control channel, 110, 200
cross-layer optimisation, 128

disruption tolerant networking, 31
distributed decision-making, 134
DSP, 172
dynamic range, 183
dynamic spectrum access, 10, 201
   overlay, 11
   underlay, 11

EIRP mask, 205
etiquette, 16
ETSI, 219

FPGA, 171

game theory, 136
GPP, 173
GPU, 173

hardware, 167
heuristics, 132
hidden node, 107

IEEE 802.22, 113–115, 163, 217
interference, 48
   adjacent channel interference, 52, 68
   c-channel interference, 53
   intermodulation distortion, 53, 185
   IP3, 185
   propagation issues, 49
   transmitter effects, 48
ITU, 217

just-in-time spectrum, 15, 208

knobs and meters, 41
knowledge representation, 141

learning, 146
   neural networks, 151
   pattern recognition, 150
   reinforcement learning, 149
   supervised learning, 149
   unsupervised learning, 149

MAC, 79, 128, 168
making decisions, 44, 123, 225
making observations, 43, 223
MANETS, 20
metaheuristics, 132
meters, 43
military, 22
MIMO, 76, 80
Mitola, 1
multi-carrier techniques, 57
multi-core, 177
   cell, 178
   DSP, 179
   GPP, 177
   picoArray, 179
multilateral decision-making, 134
multipath, 51
multiple access, 60

noise figure, 184
noise floor, 184

observe, decide, act cycle, 35, 39
OFDM, 58, 66, 69, 74
ontologies, 142, 145
optimisation, 125
  genetic algorithms, 132, 133
  hill climbing, 132
  simulated annealing, 132
  tabu search, 132

P1900, 215
personalisation, 21
platforms, 226
policy-based management, 144
power flux density, 206
public safety, 24, 162, 214

Rayleigh fading, 51
reasoning, 141
  case-based, 142
  inference, 142
regulation, 146, 195, 226
RF frontend, 182
RF performance, 183
Rician fading, 51

SCC 41, 215
SDR Forum, 219
security, 44, 155, 226
  authentication, 155
  authorisation, 155
  collaboration weaknesses, 159
  encryption, 155
  IEEE 802.22, 163
  physical fakes, 157
  public safety, 162
  single points of failure, 161
selectivity, 186

self-organising applications, 18
sensitivity, 184
service network, 117
service-neutral regimes, 12
service neutrality, 204
spatial footprint, 70
spectrum commons, 15, 200, 208
spectrum management, 4, 195
  frequency allocation, 5
  frequency assignment, 7
  monitoring, 198
  spectrum assignment, 197
  spectrum authorisation, 200
  spectrum engineering, 198, 210
  spectrum monitoring, 212
  spectrum planning, 197, 198
spectrum sensing, 92
  cooperative, 108, 112
  cyclostationary feature detection, 102
  energy detection, 99
  false alarms, 93
  feature detector, 102
  interference temperature, 117
  matched filter, 100
  missed detection, 93
  non-cooperative, 105
spectrum sharing, 10
spread spectrum, 57

taking action, 41
technology-neutral regimes, 12
technology neutrality, 204
time-limited leases, 210, 212
Tragedy of the Commons, 16
TV white spaces, 26, 229–30

umbrella example, 33
UWB, 28, 57

WiMAX, 2, 35